DEFENSE, SECURITY AND STRATEGIES

SPECIALTY METAL USE BY THE DEPARTMENT OF DEFENSE

ANALYSIS AND ISSUES

DEFENSE, SECURITY AND STRATEGIES

Additional books in this series can be found on Nova's website under the Series tab.

Additional e-books in this series can be found on Nova's website under the e-book tab.

DEFENSE, SECURITY AND STRATEGIES

SPECIALTY METAL USE BY THE DEPARTMENT OF DEFENSE

ANALYSIS AND ISSUES

MACKENZIE R. BARKER
EDITOR

NOVINKA

New York

Library of Congress Cataloging-in-Publication Data

ISBN: 978-1-63482-061-5

Published by Nova Science Publishers, Inc. † New York

Contents

PREFACE

Specialty metals are essential to the manufacture and performance capabilities of certain Department of Defense (DOD) weapon system parts, due to their unique properties, such as being highly magnetic, lightweight, corrosion resistant, or having high durability. Among these metals are samarium-cobalt alloy magnets used to make radar systems, as well as titanium and certain steel alloys used to make aircraft, submarines, and tactical ground vehicles. There are few, if any, substitutes for some of these metals. This book assesses how DOD meets its needs for specialty metals parts and ensures compliance with restrictions; and DOD's process for providing national security waivers for specialty metal procurements and the extent to which it disseminates waiver information throughout the department.

Chapter 1 – Specialty metals—such as titanium, certain steel alloys, and samarium-cobalt alloy magnets—are essential to DOD weapon systems due to their unique properties, such as being highly durable. Federal statute requires specialty metals used in weapon systems to be procured from domestic sources or qualifying countries. However, the law allows DOD to waive this requirement in the interest of national security. GAO was mandated by a House report accompanying a bill for the National Defense Authorization Act (NDAA) for Fiscal Year 2014 to review DOD's compliance with specialty metals requirements. This report assesses (1) how DOD meets its needs for specialty metals parts and ensures compliance with restrictions, and (2) DOD's process for providing national security waivers for specialty metal procurements and the extent to which it disseminates waiver information throughout the department. GAO reviewed contracts, laws, regulations and DOD guidance, and analyzed a non-generalizable sample of five weapon systems as case studies based on their total 2013 acquisition costs, among other things. GAO also reviewed national security waivers DOD granted since 2009 and interviewed DOD and contractor officials.

Chapter 2 – This report examines the specialty metal clause, potential oversight issues, and options for Congress. The specialty metal clause in the Defense Federal Acquisition Regulation Supplement (DFARS) prohibits the Department of Defense (DOD) from acquiring end units or components for aircraft, missile and space systems, ships, tank and automotive items, weapon systems, or ammunition unless these items have been manufactured with specialty metals that have been melted or produced in the United States. Thousands of products used for defense, aerospace, automotive, and renewable energy technologies rely on specialty metals for which there are often few, if any, substitutes. Specialty metals covered by this provision include certain types of cobalt, nickel, steel, titanium and titanium alloys, zirconium, and zirconium base alloys. In order to preserve and protect the United States industrial base so that it could meet DOD requirements during periods of adversity and war, Congress passed a set of domestic source restrictions which became known as the Berry Amendment. In 1973, specialty metals become one of the items covered under the Berry Amendment. Over three decades later, specialty metals are now covered in a separate citation in the United States Code (U.S.C.). Congress took action in the FY2007 National Defense Authorization Act, P.L. 109-364, to separate specialty metal from the Berry Amendment (Title 10, U.S.C. 2533a). Specialty metal provisions underwent a substantial revision in P.L. 110-181 as part of Congress's continuing effort to create a procurement environment that promotes efficiency in the DOD acquisition process, while insuring that the United States has a vigorous domestic metals industry capable of meeting defense needs. The revised specialty metal clause made clear the requirement that specific defense articles must be produced using domestic specialty metals; made exemptions for commercial-off-the-shelf (COTS) articles, electronic articles, and articles containing small amounts of non-compliant specialty metals; and allowed producers of commercially derivative defense articles to treat domestic and foreign specialty metals as fungible materials so that commercial and defense articles may be produced on the same production line without the need to trace the small amounts of metal used in each article. These changes reflected a view in Congress that there are differing rationales for offering domestic source provisions, and that these refinements would promote efficiencies throughout the defense supply chain. In October 2013, DOD released its Annual Industrial Capabilities Report to Congress in accordance with Section 2504 of Title 10, United States Code (U.S.C.). The report states that, due to global market forces, the overall demand for rare earth materials has decreased, as prices for most rare earth oxides and metals have declined since 2011. There are at least seven possible options for policymakers to consider: (1) eliminate the specialty metal clause; (2) require an assessment of compliance with new exceptions to the specialty metals clause; (3) require a review of waivers issued under the revised specialty metals clause, including requiring DOD to publicly disclose when waivers are granted; (4) require congressional approval before non-compliant specialty metal can be used in certain defense contracts; (5) require a congressional report for each platform/component where non-compliant specialty metals are used in defense contracts; (6) encourage the use of domestic specialty metal; and (7) appoint a special metals protection board.

In: Specialty Metal Use by the Department … ISBN: 978-1-63482-061-5
Editor: Mackenzie R. Barker

Chapter 1

SPECIALTY METALS: DOD DISSEMINATION OF NATIONAL SECURITY WAIVER INFORMATION COULD ENHANCE AWARENESS AND COMPLIANCE WITH RESTRICTIONS*

United States Government Accountability Office

ABBREVIATIONS

DCMA	Defense Contract Management Agency
DFARS	Department of Defense Federal Acquisition Regulation Supplement
DOD	Department of Defense
FAR	Federal Acquisition Regulation
G/ATOR	Ground/Air Task Oriented Radar
USD AT&L	Under Secretary of Defense for Acquisition, Technology and Logistics

* This is an edited, reformatted and augmented version of a report, GAO-15-133, issued by the Government Accountability Office, dated October 2014.

WHY GAO DID THIS STUDY

Specialty metals—such as titanium, certain steel alloys, and samarium-cobalt alloy magnets—are essential to DOD weapon systems due to their unique properties, such as being highly durable. Federal statute requires specialty metals used in weapon systems to be procured from domestic sources or qualifying countries. However, the law allows DOD to waive this requirement in the interest of national security. GAO was mandated by a House report accompanying a bill for the National Defense Authorization Act (NDAA) for Fiscal Year 2014 to review DOD's compliance with specialty metals requirements.

This report assesses (1) how DOD meets its needs for specialty metals parts and ensures compliance with restrictions, and (2) DOD's process for providing national security waivers for specialty metal procurements and the extent to which it disseminates waiver information throughout the department. GAO reviewed contracts, laws, regulations and DOD guidance, and analyzed a non-generalizable sample of five weapon systems as case studies based on their total 2013 acquisition costs, among other things. GAO also reviewed national security waivers DOD granted since 2009 and interviewed DOD and contractor officials.

WHAT GAO RECOMMENDS

GAO recommends that DOD disseminate non-sensitive information within the department and its supplier-base on the waivers it has granted for specialty metals. DOD concurred with the recommendation and plans to publish non-sensitive information.

WHAT GAO FOUND

The Department of Defense (DOD) typically relies on its prime contractors to plan for the procurement of specialty metals and ensure compliance with specialty metals' restrictions for the five weapon systems programs that GAO reviewed. For these programs, GAO found that DOD plays a limited role—primarily monitoring the availability of specialty metals and conducting periodic reviews of prime contractor quality assurance processes. GAO also reviewed contracts for these five programs and found they contained clauses that require prime contractors to procure specialty metals in compliance with domestic source restrictions, ensure that delivered items meet contract requirements as part of quality assurance, and maintain processes for future material needs. In turn, these prime contractors told GAO that they pass down the contract requirements—including those pertaining to specialty metals—to their subcontractors and defense suppliers, and require them to follow industry standards for quality management. These standards include, among other things, testing subcontractor processes to determine if they meet contractual specifications; reviewing required supplier certifications for items delivered under the contract to confirm compliance with all identified requirements; and rating subcontractors using performance metrics. Prime contractors for these programs also told GAO they use a risk-based approach to oversee subcontractors, including those suppliers at lower tiers.

DOD recently improved its national security waiver process; but its dissemination of information contained in those waivers is limited. Since 2009—when specialty metals restrictions were changed and the exception for national security was added—DOD has granted six national security waivers to five different weapon system programs known to have procured noncompliant specialty metals. Five of the six waivers were for samarium-cobalt magnets, which were noncompliant largely due to a change in a previously allowed exception for these magnets. During its review, GAO identified weaknesses in DOD's waiver process such as not having defined procedures for requesting waivers; and in June 2014, DOD developed written guidance for program offices to follow when requesting these waivers. However, GAO also found that DOD does not have a mechanism to share information on national security waivers granted for noncompliant specialty metals. *Standards for Internal Control in the Federal Government* call for information to be recorded and communicated to management and others, including external parties who need it, such as program offices and suppliers, to help the agency achieve its goals. Disseminating non-sensitive information— including the names of programs that received waivers, sources of the noncompliant specialty metals, and corrective actions—to key stakeholders, such as DOD weapon system program offices and their defense suppliers, could help raise awareness of and compliance with the specialty metals restrictions. Moreover, greater awareness of supplier-base problems and broader dissemination of national security waiver information could assist DOD in better discovering potential vulnerabilities, such as systemic supply chain risks that could impact national security objectives.

* * *

October 16, 2014

Congressional Committees

Specialty metals are essential to the manufacture and performance capabilities of certain Department of Defense (DOD) weapon system parts, due to their unique properties, such as being highly magnetic, lightweight, corrosion resistant, or having high durability. Among these metals are samarium-cobalt alloy magnets used to make radar systems, as well as titanium and certain steel alloys used to make aircraft, submarines, and tactical ground vehicles. There are few, if any, substitutes for some of these metals. Specialty metals domestic source restrictions have been established to ensure that the United States has a vigorous domestic metals supply chain capable of meeting defense needs.[1] These restrictions require that such metals for DOD weapon systems and certain items or components be procured from domestic sources or certain qualifying countries unless an exception applies. Exceptions include those for domestic nonavailability, and in situations where acceptance of an end item is necessary to national security. For example, if a weapon program is using noncompliant specialty metals, the Secretary of Defense may issue a waiver for that program, if the Secretary determines that the acceptance of end items made with noncompliant metals are in the interest of national security.

Recent instances of DOD's noncompliance with specialty metals restrictions and the number of requests for national security waivers have raised questions about DOD's oversight of its weapon systems contractors. GAO was mandated by a committee report accompanying a bill for the National Defense Authorization Act (NDAA) for Fiscal Year 2014 to review DOD's compliance with specialty metals' requirements and its use of waivers for national security interests (national security waivers). [2] In this report, we assessed: (1) how DOD meets its needs for specialty metals parts and ensures compliance with the specialty metals restrictions, and (2) DOD's process for providing national security waivers for specialty metal procurements and the extent to which it disseminates waiver information throughout the department.

To do our work, we examined laws and regulations regarding domestic source restrictions, including the Defense Federal Acquisition Regulation Supplement (DFARS), as well as reviewed DOD guidance relating to planning for and compliance with specialty metals and related requirements of the

Federal Acquisition Regulation (FAR). We also selected a non-generalizable sample of five weapon systems as case studies on the basis of their total acquisition cost, and phase in the acquisition process as of November 2013; as well as to include representation of at least one program from each of the military services: the Joint Strike Fighter, KC-46 Tanker, DDG-51 *Arleigh Burke* Class Destroyer, *Virginia* Class Submarine, and Family of Medium Tactical Vehicles.[3] We reviewed contracts for these programs to see if they included the specialty metals restrictions and documentation provided by contractor and DOD quality assurance staff on specialty metals planning efforts and parts management planning to understand how contractors ensure parts availability and mitigate risk. We identified the extent to which DOD depends on contractors to perform these activities. Our findings from these five programs cannot be generalized to all programs, but they provide useful insights into how DOD officials and contractors work to address requirements associated with specialty metal restrictions. We also interviewed officials from DOD, including the Office of the Under Secretary of Defense for Acquisition, Technology, and Logistics (USD AT&L), the Defense Contract Management Agency (DCMA), and selected weapon system programs, as well as their prime contractors, two selected subcontractors, and specialty metals company representatives. We ascertained their roles in determining the needs for specialty metals parts and monitoring compliance with specialty metals restrictions and related requirements of the FAR and the DFARS. We also reviewed DOD's process for providing waivers for national security interests for weapon programs that procured noncompliant specialty metals, and assessed the extent to which DOD disseminates information on national security waivers it has granted against criteria in *Standards for Internal Control in the Federal Government.*[4] We identified and reviewed the national security waivers for specialty metal procurements granted by DOD since 2009, when specialty metals restrictions were changed and the exception for national security was added.[5] For instances where noncompliance was reported and a national security waiver was requested, we reviewed corresponding DOD determinations of whether the contractor's or subcontractor's noncompliance was knowing or willful. We also reviewed legislative history for DOD's authority to grant waivers. We interviewed officials representing USD AT&L to identify other relevant background, rationale, or circumstances not stated in the waiver documentation. We did not assess whether other exceptions were available under the facts and circumstances present for these waivers. Appendix I provides a more detailed description of our scope and methodology.

We performed our review from December 2013 to October 2014 in accordance with generally accepted government auditing standards. Those standards require that we plan and perform the audit to obtain sufficient, appropriate evidence to provide a reasonable basis for our findings and conclusions based on our audit objectives. We believe that the evidence obtained provides a reasonable basis for our findings and conclusions based on our audit objectives.

BACKGROUND

In 1941, Congress enacted the Berry Amendment, which required that certain items procured for defense purposes be grown or produced in the United States.[6] Specialty metals were added to the Berry Amendment in the early 1970s. The term "specialty metals" is defined to mean any of the following:

- Steel with a maximum alloy content exceeding one or more of the following limits: manganese, 1.65 percent; silicon, 0.60 percent; or copper, 0.60 percent; or containing more than 0.25 percent of any of the following elements: aluminum, chromium, cobalt, columbium, molybdenum, nickel, titanium, tungsten, or vanadium.
- Metal alloys consisting of nickel, iron-nickel, and cobalt base alloys containing a total of other alloying metals (except iron) in excess of 10 percent.
- Titanium and titanium alloys.
- Zirconium and zirconium base alloys.

Specialty metals and their potential applications in DOD weapon systems include: steel alloys such as those used for ship hulls; metal alloys consisting of nickel and iron-nickel; certain cobalt base alloys such as samarium-cobalt alloy magnets used in radars; titanium and titanium alloy used in aircraft engine parts; and zirconium and zirconium base alloys used in gas turbine engines. In 2006, Congress passed the Fiscal Year 2007 NDAA which established specialty metals restrictions separate from the Berry Amendment. The provision generally requires DOD and its contractors to procure specialty metals produced or melted in the United States unless an exception applies which then permits the specialty metals to be obtained from foreign countries.[7]

Table 1 shows the federal regulations and DOD guidance that govern the elements of planning, compliance, and domestic source restrictions when procuring specialty metals for major weapon systems.

Table 1. Federal Regulations and DOD Guidance for Planning, Compliance, and Domestic Source Restrictions for Procuring Specialty Metals for Major Weapon Systems

Elements	Description of regulation or DOD guidance
Planning	The Defense Federal Acquisition Regulation Supplement (DFARS) provision and clause on Material Management and Accounting Systems (DFARS § 242.7202 and clause 252.242-7004) require contractors to have a material management and accounting system that, among other things, reasonably forecasts material requirements. This would also include specialty metals. DOD has a parts management standard practice (MIL-STD-3018) and guide (SD-19), which call for contractors to consider how parts will be selected as they design systems early in a program's acquisition process. The guidance also calls for contractors to create parts management plans, which can help ensure that compliant specialty metals are available for future production and sustainment of the weapon system. DOD encourages program offices to use these standards in weapon system contracts.
Compliance (Quality Assurance)	The Federal Acquisition Regulation (FAR) contains policies and procedures to ensure that supplies and services acquired under Government contract conform to quality and quantity requirements. The FAR requires contracts to include inspection and other quality requirements to protect the government's interest, and specifies that nonconforming supplies or services may be rejected by the government (FAR § 46.102). This would include compliance with the specialty metals domestic source restrictions.
Domestic Source Restrictions	The specialty metals clause (DFARS 252.225-7009) requires that specialty metals procured for DOD articles must be melted or produced in the United States, its outlying areas, or in a qualifying country, unless an exception applies. Exceptions include those for domestic nonavailability and purchases of electronic components. Domestic nonavailability includes the U.S., outlying areas, or a qualifying country. DFARS 252.225-7009(b)(5).

Source: FAR, DFARS, and DOD guidance on parts management. | GAO-15-133.

DOD also has other exceptions available under the specialty metals restrictions, but they were beyond the scope of our review. As mandated, our review focused on national security waivers. Appendix II provides a list of the allowable exceptions to the specialty metals restrictions. Some of these exceptions include:

- *National Security Waivers:* When the use of noncompliant specialty metals in a weapon system is identified after it has been fabricated, the Secretary of Defense may determine in writing that acceptance of such an end item is necessary to the national security interests of the United States. The Secretary is allowed by statute to delegate this authority to the USD AT&L. DOD generally must notify Congress in advance of making these written determinations except in the case of an urgent national security requirement.
- *Domestic Non-Availability:* A domestic non-availability exception may apply if USD AT&L or the Secretary of the military service determines that compliant specialty metal of satisfactory quality, sufficient quantity, and in the required form, cannot be procured when needed at a fair and reasonable price.
- *Qualifying Countries:* This exception waives the requirement for procuring specialty metals produced in the United States if the acquisition relates to certain agreements with foreign governments, known as "qualifying countries." Under the qualifying country exception, manufacturers in these countries have greater flexibility when procuring specialty metals for DOD procurements than U.S. manufacturers. Specifically, they can procure specialty metals from any source—including non-qualifying countries—while a component manufacturer in the United States must procure specialty metals from a source in the United States or a qualifying country.

To implement the specialty metals restrictions, DOD established a clause in the DFARS. Generally DOD must include this clause in weapon system solicitations and contracts to require that contractors deliver items incorporating specialty metals in compliance with statute and regulation. Contractors are required to insert this clause, known as flow down, in their subcontracts that include items containing specialty metals, to the extent necessary to ensure compliance. In fiscal year 2009, changes were made to the specialty metals restrictions statute by the Fiscal Year 2008 NDAA, which eliminated an exception used to procure electronic components containing high performance magnets, with noncompliant metals from non-domestic sources. It also established a national security waiver exception—permitting the Secretary of Defense to accept delivery of an end item containing non-domestic specialty metals when the Secretary certifies that it is in the interests of national security.

DOD PRIMARILY RELIES ON CONTRACTORS TO DETERMINE NEEDS FOR SPECIALTY METALS AND VERIFY COMPLIANCE WITH RESTRICTIONS

The Department of Defense (DOD) monitors the availability of specialty metals and conducts periodic quality assurance reviews; but played a limited role in planning for the procurement of specialty metals and ensuring compliance with specialty metals' restrictions for the five programs we reviewed. Officials representing the weapon system programs that we reviewed typically rely on their prime contractors to plan for the procurement of specialty metals and to ensure compliance with these restrictions. We reviewed selected prime contracts for these programs and found they contained clauses requiring prime contractors to procure specialty metals in compliance with domestic source restrictions, ensure that delivered items meet contract technical requirements as part of quality assurance, and maintain processes for meeting future material needs. In turn, these prime contractors told us that they flow down the contract requirements—including those pertaining to specialty metals—to their suppliers, and require them to follow industry standards for quality management. Further, prime contractors for these programs told us they use a risk-based approach to oversee subcontractors, including those at lower tiers. DOD quality assurance staff also conduct periodic quality assurance reviews, primarily at the prime contractor level.

DOD Relies on Weapon System Contractors and Subcontractors to Plan for Specialty Metal Needs

DOD played a limited role in planning for specialty metals' needs for the five programs we reviewed, and does so primarily through reports and assessments of availability of supply. Specifically, the Defense Logistics Agency conducts periodic assessments of strategic and critical materials for DOD, which help inform the department's decisions to undertake risk mitigations such as stockpiling materials subject to potential shortfalls.[8] In addition, the U.S. Geological Survey provides information to DOD on a variety of mineral commodities, which DOD uses in its analysis of materials availability; it also produces publicly available data on production and trends for some specialty metals and alloys used in specialty metals, including titanium. Specifically, the U.S. Geological Survey reported the price of titanium ingot increased by 250 percent between 2003 and 2006, before dropping to its average price range in the following 4 years.

Due in part to the limits of worldwide production capacity, some specialty metals require a long lead time to produce, and specific grades of titanium procured for weapon system production may also require qualification and testing periods. Growing tensions in countries that the United States depends on for these metals, such as Russia, could potentially lead to additional availability risks. For example, a specialty metals industry official told us that samarium cobalt magnets are uniquely developed for DOD weapon systems and must undergo lengthy qualification and testing periods before their use in a weapon system. Further, in DOD's most recent annual assessment of industrial capabilities, it stated that the industrial base upon which DOD relies has steadily become more global and diverse, and DOD does not control the supply chain that supports production.[9]

DOD officials from the five weapon system programs we selected for our review reported that they task their contractors to determine their respective needs for specialty metals and to plan to procure these metals in accordance with the domestic source restrictions. Beyond requiring contractors to have a material management and accounting system that reasonably forecasts material requirements, such as specialty metals, DOD has a parts management standard practice and guide which call for contractors to consider how parts—including those with specialty metals—will be selected as they design systems early in a program's acquisition process.[10] Specifically, five of the six weapon system program prime contracts we selected for review require their prime contractors to have a material management and accounting system to forecast material requirements, which when cited, applies to all materials, including specialty metals.[11] While the current Family of Medium Tactical Vehicles prime contract does not contain a material management and accounting system requirement, the current prime contractor reported that it conducted parts management planning according to DOD's parts management guidance, and the prime contractor told us that this planning includes specialty metals.

While neither the material management contract requirement for the programs in our review or the Family of Medium Tactical Vehicles program's parts management contract requirement specifically discuss specialty metals, they both require contractors to do advance planning for the materials needed in future production, which would include specialty metals or parts containing them. The six prime contractors for the programs we reviewed reported that they conduct various activities to ensure future specialty metals parts for production will be available, including forecasts of specialty metals needed for future production and initiating advance purchasing agreements with specialty metals producers. Four of the prime contractors indicated they have had no difficulties in obtaining specialty metals needed for production from domestic sources or qualifying countries, but two reported encountering availability issues for some specialty metals, including titanium pipe, stainless steel, and some engine parts metals. Table 2 summarizes the six prime contractors' planning activities for procuring specialty metals and the extent to which they have incurred availability issues.

Table 2. Specialty Metals Availability Issues and Planning Activities Reported by Selected Prime Contractors

Program	Availability	Prime contractor planning for specialty metals procurements
Virginia Class Submarine	No availability issues identified by contractor	Prime contractor reports forecasting materials needed for submarines under contract. Prime contractor representatives stated that production forecasts cover the next five to six years. The contractor reports that it has an advance purchasing agreement for steel. The prime contract requires a material management and accounting system. While the contract does not require a parts management plan, the prime contractor told us they use such a plan and that it addresses all parts, including those with specialty metals.
Joint Strike Fighter	Joint Strike Fighter	Prime contractor reports conducting periodic industry studies to determine the forecasted capacity and market conditions for titanium and reviews aerospace and metals industry trends to forecast future over and under capacity conditions. Industry suppliers provide periodic briefings. Titanium is forecast for the entire program schedule; five year forecasts are provided by aircraft. Stainless steel usage is not forecasted, but a rough basis of estimate is provided to suppliers for planning in the near term. Prime contract requires a material management and accounting system. The contractor

Table 2. (Continued)

Program	Availability	Prime contractor planning for specialty metals procurements
		has a contractual agreement with DCMA requiring a parts management plan; however this plan is used for managing standard parts, including fasteners and common hardware, and does not address specialty metals requirements. Specialty metals are instead addressed through the efforts listed above and through purchasing documents that cite the domestic source restriction as a requirement.
KC-46 Tanker	No availability issues identified by contractor	Prime contractor reports creating a five year forecast for specialty metals that covers all prime contractor-produced aircraft currently under proposal or contract with the U.S. government. The forecasts cover prime contractor requirements and a percentage of supplier requirements. This forecast utilizes estimated weights of specialty metals for each U.S. Government aircraft model manufactured by the prime contractor, as well as the equivalent commercial model. Prime contract requires a material management and accounting system. The prime contract does not require a parts management plan and the contractor does not otherwise have such a plan related to specialty metals, but instead uses its raw material procurement process to ensure specialty metals requirements are addressed.
Family of Medium Tactical Vehicles	Availability issues (types of engine parts)	Prime contractor reports reviewing materials for any industry shortages and if one exists, prime contractor reports that it will work with engineering and supply chain staff to mitigate the risk through resourcing or through the use of alternative sources, if applicable. The prime contractor creates forecasts that typically do not exceed two years. Parts management is required by the prime contract, and the prime contractor states this plan is aligned with its purchasing procedures.
DDG-51 Destroyer (the first of two prime contractors)	Availability issues (small diameter titanium pipe and types of stainless steel)	Prime contractor reports having near-term agreements with mills and distributors, based on three year forecasts for some materials that contain specialty metals, such as steel plate, fasteners, stainless pipe and tube, and carbon steel shapes. Prime contract requires a material management and accounting system. The prime contractor reports it

Program	Availability	Prime contractor planning for specialty metals procurements
		is not contractually required to do parts management planning, but does have a parts management plan and implements it by flowing down specialty metals requirements to suppliers in purchase order descriptions and quality assurance requirements.
DDG-51 Destroyer (the second of two prime contractors)	No availability issues identified by contractor	Prime contractor reports forecasting all of its requirements via its material requirements planning system and has long-term agreements for some materials, including alloy steel plate. According to the prime contractor, these agreements typically last five-to-ten years and cover over 95 percent of prime contractor's requirements, but do not include subcontractors' requirements; and subcontractors are responsible for their own specialty metals procurements. Prime contract requires a material management and accounting system. Although not required by the prime contract, the prime contractor reports that it conducts parts management planning, which includes flowing down material requirements to suppliers as part of long-term agreements with established lead-times.

Source: GAO presentation of information provided by DOD prime contractors. | GAO-15-133.

Note: We did not assess the quality of contractor or subcontractor planning efforts or their compliance with contractual requirements or the DFARS.

DOD Relies on Weapon System Contractors and Suppliers to Ensure Compliance with the Specialty Metals Restrictions, but Periodically Reviews Prime Contractor Quality Assurance Processes

DOD officials for the five weapon systems programs we selected for review reported that they contractually require their prime contractors to comply with the specialty metals restrictions. Five of the six prime contractors, in turn, reported that they rely on their subcontractors for compliance by including the specialty metals restrictions in their subcontracts and purchase orders to the extent necessary to ensure compliance of the end products delivered to the government. The other prime contractor, responsible for the development of the KC-46 Tanker, said it directly procures specialty metals for this military aircraft, rather than relying on subcontractors, largely based on the existing design for its commercially available aircraft.

In addition, we found that the prime contracts for four of the five DOD weapon system programs we selected for review require the prime contractor to adopt and use industry standards for quality management. For the remaining contract, although it is not specifically stated in the contract, the prime contractor indicated to us that they interpret their contract to require them to adopt and use industry standards for quality management. These prime contractors indicated they also require their subcontractors and suppliers to use these industry standards.[12] These standards apply to the procurement of all parts, including specialty metals, and include: (1) evaluating potential subcontractors for inclusion on the contractor's approved suppliers list; (2) reviewing required independent certifications or subcontractor certificates of conformance for items delivered under contract;[13] (3) testing subcontractor parts and processes to determine if they meet contractual specifications; and (4) rating subcontractors on a routine basis using performance metrics such as product quality and on-time deliveries.[14]

Five of the six prime contractors that we spoke with reported that these industry standards for quality management were included in their contracts. For example, the prime contractor for the Family of Medium Tactical Vehicles reported that it performs systematic quality reviews of its subcontractors every 6 months to 2 years. As part of its review, it requires suppliers to provide certificates of conformance for parts procured and reviews material certifications from all first tier suppliers and some second tier suppliers to ensure industry quality standards are met. On the other hand, the prime contractor for the KC-46 Tanker reported that it directly handles procurement of specialty metals, rather than relying on subcontractors to ensure compliance, although it also requires its suppliers to adhere to industry quality standards. In addition, each of the six prime contractors in our review reported that they use a risk-based approach—based on factors such as product complexity and the subcontractor's past performance—to determine the extent to which they conduct subcontractor oversight.

DOD also oversees prime contractors' purchasing processes and quality assurance activities, which can include reviews of whether prime contractors are complying with the specialty metals restrictions. These oversight activities are carried out by staff from DCMA—for the Army and Air Force—and from the staff of the U.S. Navy Supervisors of Shipbuilding, Conversion, and Repair (Navy) organizations. The FAR requires DOD contracting officers to determine whether contractor purchasing system reviews are needed, and if so, to conduct a review with the objective of evaluating the efficiency and effectiveness with which the contractor spends government funds and compliance with government policy when subcontracting. The DFARS also requires contracting officers to evaluate whether the contractor's purchasing system is capable of ensuring that all applicable contractor purchase orders and subcontracts contain all terms, conditions, DFARS clauses— including the flow down of the specialty metals' restriction—and any other clauses needed to carry out the requirements of the prime contract.[15] This review provides a basis for granting, withholding, or withdrawing approval of the contractor's purchasing system. After an initial review, contracting officers are to determine, at least every 3 years, whether subsequent reviews are required.

According to DCMA guidance on supplier risk management, DCMA is to identify risk levels for DOD suppliers, and to use these ratings in developing government contract quality surveillance plans and allocating resources to perform them. This guidance also provides that government oversight of sub-tier suppliers may occur if warranted. DCMA receives its direction and expectations for contract oversight from the DOD program office, which can include access to subcontractor invoices at any tier, if the program deems this necessary. In addition, the FAR provides that, for major system acquisitions, the contracting officer may designate certain high risk or critical subsystems or components for special surveillance.[16]

DOD and the prime contractor can also arrange for access in the event that noncompliance issues are identified to help ensure corrective actions are taken.

Moreover, the FAR allows contracting officers to accept a contractor's certification that the delivered items meet requirements rather than performing an inspection of these items.[17] However, if any delivered items are noncompliant, the FAR generally provides that any defect in items provided under a fixed-price contract is to be replaced, corrected, or repaired by the contractor, potentially at their own expense.[18] We reviewed the six prime contracts for the programs in our review and each contract contained clauses related to the correction of deficiencies in items delivered by the contractor.

DOD HAS DEFINED PROCEDURES FOR REQUESTING SPECIALTY METALS WAIVERS, BUT DISSEMINATION OF ITS WAIVER INFORMATION IS LIMITED

The specialty metals restrictions allow the Secretary of Defense or certain delegated officials to grant national security waivers, among other things, for the use of noncompliant specialty metals. Since 2009, we found DOD has granted six national security waivers permitting the use of noncompliant specialty metals on five different weapon system programs, including the Joint Strike Fighter. During our review and, in part, in response to our discussions with USD AT&L officials, DOD developed written guidance for program offices to follow when requesting national security waivers. However, DOD lacks a mechanism to share information contained in these waivers with key stakeholders, within the department and to its supplier-base, on national security waivers granted for noncompliance with the specialty metals restrictions. Without this information, DOD and its contractors and suppliers could be limited in their awareness of, and actions to mitigate, similar supply chain issues.

DOD Has Granted National Security Waivers Since 2009 and Recently Defined Procedures for Requesting Waivers

The specialty metals' restrictions provide authority to the Secretary of Defense or certain delegated officials, including USD AT&L, to waive compliance with the specialty metals restrictions. To do so, the Secretary or delegated official must determine, in writing, that acceptance of noncompliant specialty metals is necessary to the national security interests of the United States. Waivers in the interest of national security can be only be granted after noncompliance issues are identified, and only allow the department to accept the delivery of noncompliant end items that have already been procured, rather than in advance of their procurement.[19]

Since 2009, DOD granted six national security waivers for the use of noncompliant specialty metals on five different weapon system programs—five of which were granted after contractor disclosure that noncompliant specialty metals were, in some cases, contained in the end items delivered to DOD. The remaining waiver (for the Standard Missile-3 Block II-A program) was granted as a result of an international agreement that did not address specialty metal restrictions. Table 3 shows the six national security waivers approved for five weapon system programs since 2009 when the DFARS was revised to remove a previous exemption for high performing magnets, as provided by the National Defense Authorization Act for Fiscal Year 2008.

Table 3. National Security Waivers Granted by DOD for Specialty Metals Noncompliance Since 2009

DOD program and part involved	Specialty metal/material	Waiver date	Discovery of noncompliance	Cause of noncompliance
Standard Missile-3 Block II-A All Up Rounds	Titanium and other unspecified metals	August 2012	Identification not stated in waiver.	International agreement did not address specialty metals restrictions. As a result, noncompliant specialty metals were inadvertently obtained.
Joint Strike Fighter (radar)	Samarium-cobalt magnets	November 2012	Identification by subcontractor and notified prime contractor. (August 2012)	Subcontractor purchase orders did not cite the current Defense Federal Acquisition Regulation Supplement (DFARS) specialty metals clause.
F-16 Block 52 Aircraft (radar)	Samarium-cobalt magnets	March 2013	Identification of noncompliant specialty metals in Joint Strike Fighter radar led subcontractor to check other related radar programs. (December 2012)	Subcontractor purchase orders did not cite the current DFARS specialty metals clause.
B-1 Bomber Reliability and Maintainability Improvement Program (radar)	Samarium-cobalt magnets	May 2013	Identification of noncompliant specialty metals in Joint Strike Fighter radar led	Subcontractor purchase orders did not cite the current DFARS specialty metals clause.

Table 3. (Continued)

DOD program and part involved	Specialty metal/material	Waiver date	Discovery of noncompliance	Cause of noncompliance
			subcontractor to check other related radar programs. (December 2012)	
Joint Strike Fighter target assemblies (parts forpositioning external doors, nose and main landing gear)	Samarium-cobalt magnets	April 2013	Identification by prime contractor through quality assurance review of subcontractor. (February 2013)	Prime and subcontractor purchase orders did not cite the current DFARS specialty metals clause.
Ground/Air Task Oriented Radar (G/ATOR) Program (radar)[a]	Samarium-cobalt magnets	March 2014	Identification of noncompliant specialty metals in Joint Strike Fighter radar led contractor to check other related radar programs. (May 2013)	Government development contract did not cite the current DFARS specialty metals clause.

Source: GAO presentation of DOD information. | GAO-15-133.

[a] The G/ATOR program office requested a national security waiver in anticipation of the contractor not being able to deliver compliant radar components which would have led to their noncompliance with specialty metals restrictions. However, delays in the program's schedule allowed the contractor to provide compliant components before the government accepted delivery. The contractor has since requested that this national security waiver request be withdrawn.

Specifically, in 2009, DOD eliminated an exception in the DFARS that allowed electronic components containing noncompliant high performance (samarium-cobalt) magnets to be procured from non-domestic sources.[20] However, the companies we spoke with that were involved in requesting the national security waivers indicated they had not updated their purchasing processes to reflect this DFARS change, resulting in noncompliance with specialty metals restrictions. As a result, five of the six national security waivers were granted due to noncompliant samarium-cobalt magnets procured by three companies. For each of the five programs that procured noncompliant samarium-cobalt magnets, their national security waivers stated that these magnets met the necessary performance capability requirements for the weapon systems for which they were procured. The time frame for the contractor's discovery of noncompliant specialty metals to when the waiver was approved by DOD, ranged from two months for the Joint Strike Fighter parts for positioning external doors and nose and main landing gear, to 10 months for the Ground/Air Task Oriented Radar program. According to the specialty metal statute, weapon system contractors and subcontractors that are noncompliant with the specialty metals restrictions and receive a national security waiver must develop and implement an effective plan to ensure future compliance, also known as a corrective action plan. The prime contractor for the Joint Strike Fighter program told us that it has, as well as its subcontractors, submitted corrective action plans. In addition, the prime contractors for the F-16 Block 52 aircraft and the B-1 Bomber Reliability and Maintainability Improvement Program have also submitted a corrective action plan. There is no plan for the Ground/Air Task Oriented Radar, as a result of the prime contractor correcting the noncompliance before the program accepted delivery of the radar systems regarding the need for the waiver. DOD did not also require a corrective action plan for the Standard Missile-3 Block II-A, because the noncompliant components were provided under the terms of a bilateral agreement and were not procured by a U.S. prime contractor. The waiver stated that the specialty metals restrictions still apply because the specialty metals' components are to be integrated with U.S.-acquired missile components.

In addition, in our review of national security waivers approved by DOD, we found that these waivers contained the elements required by the specialty metals restrictions. Specifically, these included (1) written determinations stating that accepting delivery of an end item containing noncompliant materials is necessary to U.S. national security interests; (2) approval by the USD AT&L or a higher-ranking official; and (3) statement indicating the quantity of end items and the time period to which the waiver applies. However, we found that DOD's process for granting national security waivers for specialty metals had some weaknesses. For example, a DOD program office that was in the final steps of submitting a national security waiver told us they had difficulty in determining what documentation to include with their waiver request. We also found that the DFARS guidance at the time that the existing waivers were submitted did not specify the types of documents required from the requesting program office to support a waiver request. In June 2014, USD AT&L developed written guidance for program offices to include how and when the noncompliance was discovered, a complete description of all of the items or systems containing noncompliant specialty metals, the manufacturer and country of origin of the noncompliant material—if known, and estimated cost and schedule estimates to replace the noncompliant parts if a national security waiver is not granted. This DOD guidance also requires the disclosure of whether the specialty metals DFARS clause was flowed down to subcontractors and whether safety and operational implications exist.[21] In addition, to prevent misinterpretation of the current flow down requirement, DOD recently proposed amending the specialty metals DFARS clause to clarify the current requirement that the specialty metals restrictions be flowed down in subcontracts.[22] The comment date for this proposed change to the DFARS ended in August 2014, and if finalized, the clause would apply to DOD contractors.

Further, DOD has been directed by a House report accompanying H.R. 4435, the Howard P. "Buck" McKeon National Defense Authorization Act for Fiscal Year 2015, to report on the department's use of national security waivers, including documentation for all of the waivers issued since 2008 and information regarding the procedures used to issue these waivers, by December 1, 2014. [23] This report directs the Secretary of Defense to report on DOD's procedures to determine whether (1) to issue a national security waiver; (2) a supply deficiency is best addressed through the national security waiver or through the availability exception; (3) noncompliance by contractors and subcontractors is "knowing or willful," and (4) further action by DOD is necessary to prevent the recurrence of the supply chain issue that led to noncompliance and the subsequent issuance of a national security waiver. The report also directed the Secretary of Defense to report on the procedures used by DOD to monitor contractor compliance. According to a senior USD AT&L official, DOD has begun work to fulfill this reporting requirement.

DOD Lacks a Mechanism to Disseminate Information on National Security Waivers Granted

In our review of the six national security waivers granted by DOD, we found that these waivers include details, such as the company's plan for procuring specialty metals that comply with the restrictions, the production units that will be compliant in the future, and estimated time frames to re-qualify suppliers and retest equipment. DOD may also require the prime contractor to pay associated costs for noncompliance with the specialty metals restrictions related to delivered items. When this occurs, the program office's contracting officer determines the amount that is appropriate to be paid based on the nature and scope of the noncompliance. Further, for the B-1 Bomber Reliability and Maintainability Improvement Program and the F-16 Block 52 Program, the waivers specify that all costs associated with non-compliance are unallowable, including any directly associated costs incurred by their contractors. Specifically, the waivers approved by USD AT&L required the contractors for these two programs to provide consideration to the Air Force for the costs that may include the design, testing, and installation of compliant components to remediate the noncompliant items and any additional costs incurred to obtain compliance for the aircraft radar systems affected. For the remaining waivers, USD AT&L's decision on whether to request contractor consideration is pending.

The specialty metals restrictions require that DOD make a determination of whether the noncompliance by the contractor or subcontractor was knowing or willful.[24] Specifically, DOD has ordered investigations of four of the six national security waivers to determine whether the noncompliance was knowing or willful, and the results of these investigations are pending. For the other two waivers, DOD did not conduct an investigation, but nonetheless determined that the noncompliance was not knowing or willful. According to officials at USD AT&L, it is in the Under Secretary's discretion to request an investigation to assist in making their determination, if the facts surrounding the noncompliance warrant it. Further, DOD can consider suspending or debarring a contractor or subcontractor whose noncompliance has been determined to be knowing or willful, until the issues that led to noncompliance have been effectively addressed. DOD has not taken this action for any of the contractors in our review. Table 4 summarizes the status of the programs' corrective action plans for their national security waivers and DOD's determination of whether the noncompliance with the specialty metals restrictions was knowing or willful.

The specialty metals restrictions further require DOD to provide advance congressional notification in the form of a written determination to the defense committees before executing a national security waiver determination.[25] However, DOD did not notify Congress for one of the six waivers and provided little advance notice for another program.

Table 4. Status of Corrective Action Plans and DOD's Investigations and Determinations for Programs that Received National Security Waivers

DOD program with waiver	Corrective action plan	Investigation ordered	Determination of whether noncompliance was knowing or willful
Standard Missile-3 Block II-A All Up Rounds	Not required	No	Not knowing or willful
Joint Strike Fighter (radar)	Yes	No	Not knowing or willful
F-16 Block 52 Aircraft (radar)	Yes	Yes – results pending	Pending
B-1 Bomber Reliability and Maintainability Improvement Program (radar)	Yes	Yes – results pending	Pending
Joint Strike Fighter target assemblies (parts for positioning external doors, nose and main landing gear)	Yes	Yes – results pending	Pending
Ground/Air Task Oriented Radar (G/ATOR)Program (radar)	Not required	Yes – results pending	Pending

Source: GAO presentation of DOD information. | GAO-15-133.

Specifically, we found that DOD did not notify Congress for the Standard Missile-3 Block II-A program waiver and the B-1 Bomber Reliability and Maintainability Improvement Program notified Congress on the same day the waiver was signed.[26] According to USD AT&L officials, the delays in notifying Congress of the approved national security waivers, in these instances, were due to the lack of a central focal point for the waivers. As a result, USD AT&L has recently centralized the processing of all waiver requests within DOD.

Standards for Internal Control in the Federal Government call for information and communications to be recorded and communicated to others, such as stakeholders who need it, to help the agency achieve its goals.[27] However, we found that DOD lacks a mechanism to share information with key stakeholders, within the department and to its supplier base, on national security waivers granted for noncompliance with the specialty metals restrictions. Sharing information on specialty metal waivers with these key stakeholders could heighten the awareness of other programs and their suppliers who are not directly involved with the waiver of the risks as well as the consequences, and to look for similar issues in their own programs. Moreover, other program offices that work with the same supplier-base could benefit from this information. For example, in the case of the Joint Strike Fighter program's noncompliant samarium cobalt, DCMA created a notification report in February 2013. According to a deputy director within DCMA's Industrial Analysis Center, reports such as this were possibly disseminated throughout DCMA, and he stated DCMA staff may have shared these reports with the program offices to which they were assigned. In November 2013, DOD discontinued these reports. Continued disseminations of this type of information and sharing them among the DOD weapon system programs and their supplier-base could heighten awareness, potentially averting future noncompliance with the specialty metals' restriction. Further, GAO's prior work in October 2008 concluded that DOD often becomes aware of supplier base problems through informal channels, and greater visibility of supply chain issues or vulnerabilities could contribute to more formal mechanisms for addressing supply chain risks.[28] We recommended that DOD create and disseminate written requirements for reporting potential concerns about supplier-base gaps. DOD addressed this recommendation, and in its most recent instruction on Defense Industrial Base Assessments, dated July 2014, it states that if an industrial capability is identified as endangered, DOD components must, among other things, validate whether the industrial capability is relevant to a satisfaction of a national security requirement. This instruction also states that to facilitate efficient and effective sharing, a repository of reports, information, and data will be established by DOD and will contain a searchable index of reports. Moreover, DOD's 2014 Chief Freedom of Information Act Officer's Report to the Department of Justice states that the department continues to implement open government principles as defined by an Office of Management and Budget memorandum, which encourages greater transparency, participation and collaboration by publishing non-sensitive government information online for public review.[29] Establishing a mechanism to disseminate non-sensitive information—including the names of programs that received waivers, sources of the noncompliant specialty metals, and corrective actions—to key stakeholders, such as DOD weapon system program offices and their defense suppliers—could help raise awareness of and greater compliance with the specialty metals restrictions.

CONCLUSION

DOD relies on its prime contractors and subcontractors to plan for and ensure that specialty metals procured for weapon systems meet the requirements of the specialty metal statutory provisions and contractor self-disclosure is the primary way that DOD becomes aware of noncompliance with this statutory provision. Some recent breaches of domestic source restrictions have led contractors to notify the government of noncompliance issues and since these instances occurred, DOD has granted national security waivers for the affected programs. DOD has recently defined procedures for requesting national security waivers for weapon systems programs. However, DOD's lack of a mechanism to disseminate information to its key stakeholders within the department's weapon system program offices and their supplier-base on national security waivers granted makes it difficult for others not directly involved with the waiver request to have awareness of the risks as well as the consequences. Further, greater awareness of supplier-base problems and broader dissemination of information could assist DOD in better discovering vulnerabilities, such as systemic supply chain risks that affect national security objectives.

RECOMMENDATION FOR EXECUTIVE ACTION

To provide greater awareness of and compliance with the specialty metal restrictions among DOD weapon system programs and their defense supplier-base, we recommend that the Secretary of Defense establish a mechanism for sharing and distributing non-sensitive information about national security waivers throughout the department and the defense supplier-base.

AGENCY COMMENTS AND OUR EVALUATION

DOD provided us with written comments on a draft of this report. In its comments, DOD concurred with our recommendation to provide greater awareness of and compliance with the specialty metal restrictions, stating that it will post non-sensitive specialty metals national security waiver information on the Defense Procurement and Acquisition Policy website. This publicly available information would include the number of national security waivers approved by DOD since 2009; the names of DOD weapon system programs receiving them; guidance on how to request a national security waiver, as well as guidance for applying specialty metals' restrictions to high performance samarium-cobalt magnets. We believe that this is a positive step towards enhancing the public transparency of future national security waivers for non-compliant specialty metals. DOD is considering resuming the issuance of notification reports when a specialty metals' noncompliance has occurred. We believe that disseminating this information across DOD could help officials working on other DOD programs assess whether similar non-compliance issues may affect them. We also received technical comments from the Department of the Interior and the DOD prime contractors included in our review, which we incorporated into this report as appropriate.

Marie A. Mak
Acting Director
Acquisition and Sourcing Managemen

APPENDIX I: OBJECTIVES, SCOPE AND METHODOLOGY

This report focuses on planning, compliance, and national security waivers for specialty metals used in Department of Defense (DOD) weapon systems. Specifically, we assessed (1) how DOD meets its needs for specialty metals parts and ensures compliance with the specialty metals restrictions, and (2) DOD's process for providing national security waivers for specialty metal procurements and the extent to which it disseminates waiver information throughout the department.

To assess how DOD officials meet their needs for specialty metals used in DOD weapon systems and ensure compliance with specialty metals restrictions, we examined laws and regulations regarding specialty metals domestic source restrictions, including the Defense Federal Acquisition Regulation Supplement (DFARS) and DOD guidance relating to planning for and compliance with specialty metals and related requirements of the Federal Acquisition Regulation (FAR). We reviewed DOD guidance on planning for weapon systems procurements and manufacturing; guidance on purchasing system reviews and quality assurance from the Defense Contract Management Agency and U.S. Navy Supervisors of Shipbuilding, Conversion and Repair; Defense Logistics Agency's reports on strategic and critical materials that identified materials with projected availability risks; and U.S. Geological Survey's price information for titanium to identify historical fluctuations in pricing and availability for specialty metals. We also interviewed officials from DOD, including the Office of the Under Secretary of Defense for Acquisition, Technology, and Logistics (USD AT&L), the Defense Contract Management Agency (DCMA), and selected weapon system programs, as well as their prime contractors, two selected subcontractors, and specialty metals company representatives. We ascertained their roles in determining the needs for specialty metals parts and monitoring compliance with specialty metals restrictions and related requirements of the FAR and the DFARS. We also interviewed and obtained written responses to questions on specialty metals planning from six prime contractors responsible for five major DOD weapon systems programs,. The programs we selected were: the F35 Joint Strike Fighter (Joint Strike Fighter), KC-46 Tanker, DDG-51 Destroyer, *Virginia* Class Submarine, and Family of Medium Tactical Vehicles, to identify how these programs plan for specialty metals needs and ensure contract requirements are met. We selected a non-generalizable sample consisting of these five DOD weapon system programs, based on their greatest total acquisition cost estimates as of November 2013. Our selection process also ensured that at least one program was selected from among each of the military services and from the following four acquisition lifecycle phases: (1) between development start and production start (e.g., the U.S. Air Force's KC-46 Tanker); (2) between production start and initial capability (e.g., DOD's Joint Strike Fighter); (3) in production and passed initial capability (e.g., the U.S. Navy's DDG-51 Destroyer and *Virginia* Class Submarine); and (4) nearing the end of production (e.g., the U.S. Army's Family of Medium Tactical Vehicles). We interviewed and obtained written responses to questions and related contractor documents from these prime contractors regarding specialty metals availability, planning activities on these programs, and the methods they use to ensure quality assurance and oversee subcontractors and suppliers. Our findings from these five programs cannot be generalized to all programs, but they provide useful insights into how DOD officials and contractors work to address requirements associated with specialty metal restrictions.

In reviewing these programs, we also obtained answers to written questions regarding specialty metals planning and quality assurance from

DOD quality assurance staff overseeing these programs, including the Defense Contract Management Agency (DCMA) and the Navy. We reviewed the results of DOD's purchasing system reviews for the six prime contractors in our review to determine whether specialty metals requirements were flowed down to subcontractors consistent with DFARS. We also reviewed written responses to questions regarding how these agencies oversee the prime contractors to help ensure compliance with their government contracts, including specialty metals domestic source restrictions. Further, we reviewed the contracts to determine what requirements related to specialty metals planning and quality assurance were included, as well as the exceptions to domestic source restrictions applicable to some of these programs, such as the KC-46 Tanker's commercial derivative military article exception. We obtained updated information regarding these methods from the prime contractors included in our review.

We reviewed and analyzed the six national security waivers that the USD AT&L approved from fiscal years 2009 through 2014. These six national security waivers included the B-1 Bomber Reliability and Maintainability Improvement Program, F-16 Block 52 Aircraft, Ground/Air Task Oriented Radar, Joint Strike Fighter radar and target assemblies parts for positioning external doors, nose and main landing gear, and Standard Missile-3 Block II-A All Up Rounds. We also assessed whether these waivers contained required elements, including whether Congress was provided advance notifications of the national security waiver determinations before they were executed as required in accordance with the statue and regulation. In addition, we obtained documents and interviewed officials from DOD's Defense Procurement and Acquisition Policy office to review their process for granting national security waivers, including how they make determinations to grant the six waivers for national security and the extent to which DOD disseminates this information throughout the department. We assessed the extent to which DOD disseminates information on national security waivers it has granted consistent with criteria in *Standards for Internal Control in the Federal Government.*[1] We reviewed prime contractor and subcontractor corrective action plans that were submitted to DCMA regarding noncompliant specialty metals and also examined the congressional notification letters relating to national security waivers that DOD provided to congressional defense committees. We met with officials at the C-130 program office at Wright-Patterson Air Force Base, in Dayton, Ohio to discuss possible plans to request a national security waiver and to discuss their experience in requesting a national security waiver. We did not assess whether other exceptions to the specialty metals restrictions were available under the facts and circumstances present for these waivers we reviewed.

We conducted this performance audit from December 2013 to October 2014 in accordance with generally accepted government auditing standards. Those standards require that we plan and perform the audit to obtain sufficient, appropriate evidence to provide a reasonable basis for our findings and conclusions based on our audit objectives. We believe that the evidence obtained provides a reasonable basis for our findings based on our audit objectives.

APPENDIX II: EXCEPTIONS TO THE SPECIALTY METALS RESTRICTION REQUIREMENTS AND QUALIFYING COUNTRY EXCEPTION

The specialty metals domestic source restrictions (10 U.S.C. § 2533b), summarized below, provide exceptions for:

- Waiver for national security interests;
- Compliant specialty metals are not available in satisfactory quality and sufficient quantity, in the required form, and cannot be procured when needed;
- Acquisitions made outside of the United States in support of combat or contingency operations;
- The use of other than competitive procedures, in accordance with the Competition in Contracting Act (10 U.S.C. 2304(c)(2) for circumstances of unusual and compelling urgency of need;
- Compliance with agreements with foreign governments;
- Commissaries, exchanges, and other non-appropriated fund instrumentalities;
- Small purchases (below the simplified acquisition threshold);
- Electronic components;
- Acquisition of some commercial items;
- Acquisition of certain commercial-off-the-shelf items;
- Acquisition of components if there is less than 2 percent of noncompliant metal (called the “de minimis” exception);
- Acquisition of certain commercially derivative defense articles; and

- Acquisition of certain noncompliant materials if the Secretary of Defense certifies in writing that acceptance of such materials is required for reasons of national security, including certain conditions and requirements.

In addition, the specialty metals clause (DFARS 225.252.7009) requires that specialty metals procured for Department of Defense articles must be melted or produced in the United States, its outlying areas, or a qualifying country. There are 23 qualifying countries, as shown in the table below.

Table 5. Qualifying Countries

Australia	France	Poland
Austria	Germany	Portugal
Belgium	Greece	Spain
Canada	Israel	Sweden
Czech Republic	Italy	Switzerland
Denmark	Luxembourg	Turkey
Egypt	Netherlands	United Kingdom of Great Britain and Northern Ireland.
Finland	Norway	

Source: GAO presentation of Defense Federal Acquisition Regulation Supplement (DFARS) § 225.003(10). | GAO-15-133.

End Notes

[1] The specialty metals restrictions, 10 U.S.C. § 2533b, contain numerous possible exceptions to the domestic source restrictions, such as exceptions for national security and commercial derivative military items. This report primarily focuses on the exception for national security interests. Throughout this report, we refer to this exception as the national security waiver.

[2] House of Representatives Report 113-102, accompanying H.R. 1960, 113th Cong. (1st Sess. (2013). H.R. 3304 was enacted as the NDAA for Fiscal Year 2014, Pub. L. No. 113- 66 (2013).

[3] The programs represent the following four acquisition lifecycle phases: (1) between development start and production start (e.g., the U.S. Air Force's KC-46 Tanker); (2) between production start and initial capability (e.g. DOD's Joint Strike Fighter); (3) in production and passed initial capability (e.g., the U.S. Navy's DDG-51 Destroyer and Virginia Class Submarine); and (4) nearing the end of production (e.g., the U.S. Army's Family of Medium Tactical Vehicles).

[4] GAO: Standards for Internal Control in the Federal Government, GAO/AIMD-00-21.3.1, Washington, D.C.; Nov. 1999.

[5] In January 2008, the National Defense Authorization Act for Fiscal Year 2008 amended 10 U.S.C. § 2533b to eliminate the use of an exception that allowed procurement of electronic components containing de minimis specialty metals from non-domestic sources, which was interpreted to permit procurement of electronic components containing high performance magnets with specialty metals from non-domestic sources. It also added the exception for national security waivers. Pub. L. No. 110-181 § 804. DOD revised the Defense Federal Acquisition Regulation Supplement in 2009 to reflect these changes. 74 Fed. Reg. 37,626 (July 29, 2009).

[6] This domestic source restriction, which became known as the Berry Amendment, initially was enacted as part of the Fifth Supplemental National Defense Appropriation Act, 1941, Pub. L. No. 77-29, 55 Stat. 123, 125 (1941). Subsequently, it was included in various defense appropriation acts and was codified in the National Defense Authorization Act for Fiscal Year 2002 at 10 U.S.C. § 2533a, Pub. L. No. 107-107 § 832 (2001).

[7] The specialty metals provision of the Berry Amendment was enacted in the Department of Defense Appropriation Act, 1973, Pub. L. No. 92-570, § 724 (1972). Although the statute did not specifically define "specialty metals," DOD defined the term consistent with the pertinent report of the House Committee on Appropriations to include four categories of metals. DOD Secretary of Defense Memorandum, "Section 724 of the Department of Defense Appropriation Act, 1973" (Nov. 20, 1972). In 2006, the specialty metals provision was codified at 10 U.S.C. § 2533b. John Warner National Defense Authorization Act for Fiscal Year 2007, Pub. L. No. 109-364, § 842 (2006).

[8] The Report to Congress on the National Defense Stockpile Requirements for strategic and critical materials is submitted to Congress by the Department of Defense pursuant to the Strategic and Critical Materials Stock Piling Act (50 U.S.C. § 98h-2).

[9] U.S. Department of Defense, Office of the Deputy Assistant Secretary of Defense for Manufacturing and Industrial Base Policy, Annual Industrial Capabilities Report to Congress (October 2013).

[10] DOD MIL-STD-3018, Standard Practice for Parts Management (Oct. 27, 2011) and SD-19, Parts Management Guide (December 2013).

[11] DOD's policy is for contractors to have a material management and accounting system that conforms to certain standards for contracts exceeding the simplified acquisition threshold which are not for commercial items and are either cost-reimbursement contracts or fixed-price contracts with progress payments. This DOD policy does not apply to small businesses. DFARS §§ 242.7200, 242.7202 and 252.242-7004.

[12] Industry quality management standards used by DOD weapon systems programs include the International Organization for Standardization's ISO 9001 standards and the Society of Automotive Engineers' AS9100 standard.

[13] A certificate of conformance is a certification, provided by a supplier confirming compliance with all identified requirements. FAR clause 52.246-15.

[14] GAO, Defense Inventory: DOD and Prime Contractors Adhered to Requirements in Selected Contracts for Overseeing Spare Parts Quality, GAO-05-73 (Washington, D.C.: Dec. 20, 2004).

[15] FAR §§ 44.301, 44.302 and DFARS § 244.305-70 and clause 252.244-7001. Generally contractors are subject to these reviews when sales to the government are expected to exceed $25 million in a 12-month period, although this excludes certain contract types, including those that are competitively awarded firm-fixed-price, fixed price with economic price adjustment and commercial item contracts.

[16] FAR §§ 42.202(e)(2) and (f).

[17] FAR § 46.504.

[18] FAR § 46.407(a) and clause 52.246-2(f).

[19] In certain cases, such as where an end item has not yet been delivered, DOD has other exceptions available under the specialty metals restrictions, but these were not in the scope of our review.

[20] In January 2008, the 2008 National Defense Authorization Act for Fiscal Year 2008 amended 10 U.S.C. § 2533b, to eliminate the use of an exception that allowed procurement of electronic components containing de minimis specialty metals from non-domestic sources, which was interpreted to permit procurement of electronic components containing high performance magnets with specialty metals from non-domestic sources. It also added the exception for national security waivers. Pub. L. No. 110-181 § 804. DOD revised the Defense Federal Acquisition Regulation Supplement in 2009 to reflect these changes. 74 Fed. Reg. 37,626 (July 29, 2009).

[21] DFARS Procedures, Guidance, and Information 22 5.7003-3(d).

[22] 79 Fed. Reg. 35,507 (June 23, 2014).

[23] H.R. Rep. No. 113-446 at 184-185 (2014) accompanying H.R.4435, Howard P. "Buck" McKeon National Defense Authorization Act for Fiscal Year 2015, 113th Cong. (2nd Sess.).

[24] 10 U.S.C. §2533b(k)(3)(A); DFARS 225.7003-3(d)(2)(ii).

[25] 10 U.S.C § 2533b(k)(2)(C); DFARS § 225.7003-3(d)(1)(iii).

[26] We recognize that neither the statute nor DFARS specifies how far in advance the notice to Congress is to be made. The statute states that the written waiver determination is to be provided to the congressional defense committees "prior to making such a determination."10 U.S.C. § 2533b(k)(2)(C). DFARS provides that the written waiver determination is to be provided "to the Congressional defense committees before the determination is executed." DFARS § 225.7003-3(d)(1)(iii). Under both the statute and DFARS in the case of an urgent national security requirement, the determination may be provided to the congressional defense committees up to 7 days after it is executed.

[27] See GAO/AIMD-00-21.3.1.

[28] GAO, Department of Defense: A Departmentwide Framework to Identify and Report Gaps in the Defense Supplier Base Is Needed, GAO-09-5 (Washington, D.C.: Oct. 7, 2008).

[29] Office of Management and Budget, Memorandum for Heads of Executive Departments and Agencies, Open Data Policy—Managing Information as an Asset (May 9, 2013).

End Note – Appendix I

[1] GAO: Standards for Internal Control in the Federal Government, GAO/AIMD-00-21.3.1 (Washington, D.C.; Nov. 1999).

In: Specialty Metal Use by the Department ... ISBN: 978-1-63482-061-5
Editor: Mackenzie R. Barker

Chapter 2

THE SPECIALTY METAL CLAUSE: OVERSIGHT ISSUES AND OPTIONS FOR CONGRESS*

Valerie Bailey Grasso

SUMMARY

This report examines the specialty metal clause, potential oversight issues, and options for Congress. The specialty metal clause in the Defense Federal Acquisition Regulation Supplement (DFARS) prohibits the Department of Defense (DOD) from acquiring end units or components for aircraft, missile and space systems, ships, tank and automotive items, weapon systems, or ammunition unless these items have been manufactured with specialty metals that have been melted or produced in the United States. Thousands of products used for defense, aerospace, automotive, and renewable energy technologies rely on specialty metals for which there are often few, if any, substitutes. Specialty metals covered by this provision include certain types of cobalt, nickel, steel, titanium and titanium alloys, zirconium, and zirconium base alloys.

In order to preserve and protect the United States industrial base so that it could meet DOD requirements during periods of adversity and war,

---

* This is an edited, reformatted and augmented version of a Congressional Research Service publication RL33751, prepared for Members and Committees of Congress dated February 6, 2014.

Congress passed a set of domestic source restrictions which became known as the Berry Amendment. In 1973, specialty metals become one of the items covered under the Berry Amendment. Over three decades later, specialty metals are now covered in a separate citation in the United States Code (U.S.C.). Congress took action in the FY2007 National Defense Authorization Act, P.L. 109-364, to separate specialty metal from the Berry Amendment (Title 10, U.S.C. 2533a).

Specialty metal provisions underwent a substantial revision in P.L. 110-181 as part of Congress's continuing effort to create a procurement environment that promotes efficiency in the DOD acquisition process, while insuring that the United States has a vigorous domestic metals industry capable of meeting defense needs. The revised specialty metal clause made clear the requirement that specific defense articles must be produced using domestic specialty metals; made exemptions for commercial-off-the-shelf (COTS) articles, electronic articles, and articles containing small amounts of non-compliant specialty metals; and allowed producers of commercially derivative defense articles to treat domestic and foreign specialty metals as fungible materials so that commercial and defense articles may be produced on the same production line without the need to trace the small amounts of metal used in each article. These changes reflected a view in Congress that there are differing rationales for offering domestic source provisions, and that these refinements would promote efficiencies throughout the defense supply chain.

In October 2013, DOD released its Annual Industrial Capabilities Report to Congress in accordance with Section 2504 of Title 10, United States Code (U.S.C.). The report states that, due to global market forces, the overall demand for rare earth materials has decreased, as prices for most rare earth oxides and metals have declined since 2011.

There are at least seven possible options for policymakers to consider: (1) eliminate the specialty metal clause; (2) require an assessment of compliance with new exceptions to the specialty metals clause; (3) require a review of waivers issued under the revised specialty metals clause, including requiring DOD to publicly disclose when waivers are granted; (4) require congressional approval before non-compliant specialty metal can be used in certain defense contracts; (5) require a congressional report for each platform/component where non-compliant specialty metals are used in defense contracts; (6) encourage the use of domestic specialty metal; and (7) appoint a special metals protection board.

INTRODUCTION

This report examines the specialty metal clause in the Defense Federal Acquisition Regulation Supplement (DFARS), potential oversight issues, and options for Congress. The specialty metal clause prohibits DOD from acquiring end units or components for aircraft, missile and space systems, ships, tank and automotive items, weapon systems, or ammunition unless these end items or components have been manufactured with specialty metals that have been melted or produced in the United States. Specialty metals are used in components procured through DOD contracts, primarily for military aircraft, weapons, equipment, and within integrated circuits, wiring, and electrical components. Specialty metals covered by this clause include certain types of cobalt, nickel, steel, titanium and titanium alloys, zirconium, and zirconium base alloys. Thousands of products used for defense, aerospace, automotive, and renewable energy technologies rely on specialty metals for which there are often few, if any, substitutes.

The availability of sources of supply of some elements that comprise some metals, particularly the access to rare earth elements, is an issue raised in recent news reports and congressional hearings. Policymakers are interested in the specialty metal clause because the specialty metal clause affects defense contractors who produce components for commercial weapons systems.

In order to preserve and protect the United States industrial base so that it could meet DOD requirements during periods of adversity and war, in 1941 Congress passed a set of domestic source restrictions which became known as the Berry Amendment.[1] In 1973, specialty metals become one of the items covered under the Berry Amendment, a domestic preference law requiring DOD to procure items that are wholly domestic. The specialty metal clause first appeared in the 1973 Defense Appropriations Act.[2] At that time, Congress was concerned with protecting domestic source materials for the Vietnam War. At that time, the domestic specialty metal sector was hurt by subsidized imports into the United States. In order to insure an adequate domestic base for domestic items, Congress provided a guarantee to domestic suppliers for a portion of DOD's specialty metal business.

Over three decades later, specialty metals are no longer part of the Berry Amendment and are now covered in a separate citation in the United States Code (U.S.C.). Congress took action in the FY2007 National Defense Authorization Act (P.L. 109-364) to separate specialty metal from the Berry Amendment (Title 10, U.S.C. 2533a). Specialty metals are defined in 10

U.S.C. 2533b and were revised as part of the FY2008 National Defense Authorization Act (P.L. 110-181).[3]

Definition of Specialty Metals

The current definition of specialty metals can be found in 10 U.S.C. 2533b, as described here:

> Specialty Metal Defined.—In this section, the term "specialty metal" means any of the following:
>
> (1) Steel—
>
> with a maximum alloy content exceeding one or more of the following limits: manganese, 1.65 percent; silicon, 0.60 percent; or copper, 0.60 percent; or
>
> containing more than 0.25 percent of any of the following elements: aluminum, chromium, cobalt, columbium, molybdenum, nickel, titanium, tungsten, or vanadium.
>
> (2) Metal alloys consisting of nickel, iron-nickel, and cobalt base alloys containing a total of other alloying metals (except iron) in excess of 10 percent.
>
> (3) Titanium and titanium alloys.
>
> (4) Zirconium and zirconium base alloys.[4]

According to DFARS 225.252.7009,[5] specialty metals procured by DOD and used in defense articles must be melted in the United States or a "qualifying country," or melted anywhere but incorporated into an article that is manufactured in a qualifying country.[6] The specialty metals clause allows a qualifying country to manufacture parts from metal that was melted anywhere, provided it meets specifications, but a United States company can only use metal that was melted in the United States or a qualifying country.

Exceptions to the Specialty Metals Clause

The specialty metal clause provides for numerous exceptions, as described here:

- Waiver for national security purposes;
- Exceptions when compliant specialty metals are not available in satisfactory quality and sufficient quantity, in the required form, and cannot be procured as and when needed;
- Exceptions for acquisitions made outside of the United States in support of combat or contingency operations;
- Exceptions for the use of other than competitive procedures, in accordance with the Competition in Contracting Act (10 U.S.C. 2304 [c]) for circumstances of unusual and compelling urgency of need;
- Exceptions to comply with agreements with foreign governments;
- Exceptions for commissaries, exchanges, and other non-appropriated fund instrumentalities;
- Exceptions for small purchases (below the simplified acquisition threshold);
- Exceptions for electrical components;
- Exceptions for the acquisition of some commercial items;
- Exceptions for the acquisition of certain commercial-off-the-shelf items;
- Exceptions for the acquisition of components if there is less than 2% of noncompliant metal (called the "de minimis" exception);
- Exceptions for the acquisition of certain commercially derivative defense articles[7]; and
- Exceptions for the acquisition of certain noncompliant materials if the Secretary of Defense certifies in writing that acceptance of such materials is required for reasons of national security, including certain conditions and requirements.[8]

MAJOR NEW DEVELOPMENTS

H.R. 1960/S. 1197, National Defense Authorization Act (NDAA) for Fiscal Year (FY) 2014

H.R. 1960 was introduced in the House on May 14, passed the House in a recorded vote (315- 108) on June 14, and was referred to the Senate on July 8, 2013.[9] S. 1197 was introduced on June 20, 2013 and referred to the Armed Services Committee. The bill contains several provisions which give the President more authority to conserve strategic and critical materials, as well as direct the Secretary of Defense to report on plans to assess the supply chain diversification for rare earth substitutes and develop risk mitigation strategies. The provisions appear below.

Subtitle B—National Defense Stockpile

Section 1411. Use of National Defense Stockpile for the Conservation of a Strategic and Critical Materials Supply

Section 1411 modifies the President's authority to maintain and manage a national defense stockpile, and allow the Defense Logistics Agency to more proactively engage in the market. These changes would grant the President the authority to conserve strategic and critical materials.

(a) Presidential Responsibility for Conservation of Stockpile Materials - Section 98e (a) of Title 50, United States Code, is amended:

by redesignating paragraphs (5) and (6) as paragraphs (6) and (7), respectively; and

by inserting after paragraph (4) the following new paragraph (5):

"(5) provide for the recovery of any strategic and critical material from excess materials made available for recovery purposes by other Federal agencies;"

(b) Uses of National Defense Stockpile Transaction Fund - Section 98h (b) (2) of Title 50, United States Code, is amended—

(1)by redesignating subparagraphs (D) through (L) as subparagraphs (E) through (M), respectively; and

(2)by inserting after subparagraph (C) the following new subparagraph (D): "(D) Encouraging the conservation of strategic and critical materials."

(c) Development of Domestic Sources - Section 98h-6(a) of Title 50, United States Code, is amended, in the matter preceding paragraph (1), by inserting 'and conservation' after 'development'.

Section 1412. Authority to Acquire Additional Materials for the National Defense Stockpile

Section 1412 provides authority to acquire certain additional strategic and critical materials for the National Defense Stockpile. The materials anticipated to be acquired have been identified to meet the military, industrial, and essential civilian needs of the United States.

(a) Acquisition Authority - Using funds available in the National Defense Stockpile Transaction Fund, the National Defense Stockpile Manager may acquire the following materials determined to be strategic and critical materials required to meet the defense, industrial, and essential civilian needs of the United States:

1) Ferroniobium.
2) Dysprosium Metal.
3) Yttrium Oxide.
4) Cadmium Zinc Tellurium Substrate Materials.
5) Lithium Ion Precursors.
6) Triamino-Trinitrobenzene and Insensitive High Explosive Molding Powders.

(b) Amount of Authority - the National Defense Stockpile Manager may use up to $41,000,000 of the National Stockpile Transaction Fund for acquisition of the materials specified in subsection (a).

(c) Fiscal Year Limitation - The authority under this section is available for purchases during Fiscal Year 2014 through Fiscal Year 2019.

Directive Report Language

In H.R. 1960, under Title XVI, Industrial Base Matters, there are two reporting requirements required by the House Armed Services Committee that address congressional concerns over maintaining a secure access and a diverse supply chain for rare earth elements to be used for national security purposes and in defense weapon systems.

The first directive requires the Under Secretary of Defense for Acquisition, Technology and Logistics to submit a report to the congressional defense committees, by February 1, 2014, to outline a risk mitigation strategy focused on securing the necessary supplies of rare earth elements. The report language reads as follows:

Title XVI – Industrial Base Matters

Report on the Diversification of Supply Activities Related to Rare Earth Elements

> The committee is aware that in response to the report required by section 843 of the Ike Skelton National Defense Authorization Act for Fiscal Year 2011 (P.L. 111-383) and based on forecasting demand for fiscal year 2013 only, the Under Secretary of Defense for Acquisition, Technology, and Logistics concluded that domestic production of rare

earth elements could satisfy the level of consumption required to meet defense procurement needs by fiscal year 2013, with the exception of yttrium. However, the committee observes that the Future Years Defense Program indicates that consumption of rare earth elements is expected to increase after 2013. Specifically, the report on the feasibility and desirability of recycling, recovery, and reprocessing of rare earth elements required by the conference report (H.Rept. 112-329) to accompany the National Defense Authorization Act for Fiscal Year 2012, states that each SSN-774 *Virginia*-class submarine would require approximately 9,200 pounds of rare earth materials, each DDG51 Aegis destroyer would require approximately 5,200 pounds of these materials, and each F-35 Lightning II aircraft would require approximately 920 pounds of these materials.

The committee is aware that the Department of Defense intends to pursue a three-pronged strategy to secure supplies of rare earth elements, which consists of diversification of supply, pursuit of substitutes, and a focus on reclamation of waste, as part of a larger U.S. Government recycling effort. The committee believes that diversification of supply activities related to rare earth elements is necessary in order to meet the growing demand for these materials, but the committee is concerned that some of these processes may prove to be technically difficult or so expensive that they are deemed cost-prohibitive.

Therefore, the committee directs the Under Secretary of Defense for Acquisition, Technology, and Logistics to submit a report to the congressional defense committees by February 1, 2014, on the Department's risk mitigation strategy for rare earth elements, which should include, at a minimum, the following elements:

1) A list and description of the programs initiated or planned to reclaim rare earth elements by the Department, along with a description of the materials reclaimed or expected to be reclaimed from such programs;
2) An assessment of the cost of materials produced by these reclamation efforts compared to the cost of newly-mined materials;
3) An assessment of availability of reliable suppliers in the National Defense Industrial Base for the reclamation and reprocessing of rare earth elements;
4) A list of alternative sources of supply, such as mine tailings, recycled components, and consumer waste, that the Department has investigated or plans to investigate;

5) A physical description of alternative sources of supply with corresponding geologic characteristics, such as grade, resource size, and the amenability of that feedstock to metallurgical processing;
6) A description of the materials that the Department plans to obtain via the Defense Priorities and Allocations System; and
7) (7) Other diversification of supply activities deemed relevant by the Under Secretary.[10]

The second directive requires DOD to perform an assessment of the potential for incorporating the substitution of non-rare earth materials into components of the Joint Strike Fighter, based on the supply chain challenges faced in securing components containing rare earth materials.

Report on the Implementation of Rare Earth Elements Strategy in the Joint Strike Fighter Program

The committee is aware that the Department of Defense intends to pursue a three-pronged strategy to secure supplies of rare earth elements, which consists of diversification of supply, pursuit of substitutes, and a focus on reclamation of waste as part of a larger U.S. Government recycling effort. However, it remains unclear how this strategy will be implemented in the Department's major defense acquisition programs (MDAPs). Several high-profile MDAPs, including the F-35 Lightening II program, may use significant amounts of rare earth elements in full-rate production. The committee is concerned that the introduction of substitute materials and components may increase acquisition and sustainment costs through the qualification of manufacturers for substitutes, implementation of engineering changes to accommodate substitutes, and the long-term costs associated with supplier networks.

Therefore, the committee directs the Assistant Secretary of the Navy for Research, Development and Acquisition, in coordination with the Program Executive Officer for the F-35, to submit a report to the congressional defense committees by February 15, 2014, on the potential for substitution of components and materials into F-35 aircraft to reduce consumption of rare earth materials. The report, which may include a classified annex, should include the following:

1) A list and description of subsystems that contain rare earth elements and the approximate quantities of each rare earth element by subsystem;

2) An assessment of the potential to incorporate substitute components or materials in each subsystem based on technical acceptability, to include consideration of performance requirements, and engineering changes that may be necessary for integration of the substitute; and
3) An assessment of the potential to incorporate substitute components or materials in each subsystem based on cost acceptability to include consideration of material costs, qualification and testing costs, and engineering change costs.[11]

S. 1600, Critical Minerals Policy Act of 2013

S. 1600, the Critical Minerals Policy Act of 2013, was introduced on October 29, 2013, and referred to the Energy and Natural Resources Committee. The bill would require the Secretary of Interior and the Secretary of Energy to amend current policies, including "facilitate the reestablishment of domestic, critical mineral designation, assessment, production, manufacturing, recycling, analysis, forecasting, workforce, education, research, and international capabilities in the United States."[12] The Senate Energy and Natural Resources Committee held a hearing on the bill on January 28, 2014. The bill has 18 sponsors and bi-partisan support.

H.R. 761, National Strategic and Critical Minerals Production Act of 2013

H.R. 761, the National Strategic and Critical Minerals Production Act of 2013 was introduced on February 15, 2013, and referred to the Committee on Natural Resources on July 8, 2013, (H.Rept. 113-138). The bill passed in a recorded vote, 246-178, and was referred to the Senate Energy and Natural Resources Committee on December 19, 2013. The bill would require both the Secretary of the Interior and the Secretary of Agriculture to more efficiently develop domestic sources of the minerals and materials of strategic and critical importance to U.S. economic and national security, and manufacturing competitiveness.[13]

H.R. 981, RARE Act of 2013

H.R. 981, the RARE Act of 2013, was introduced on March 6, 2013, referred to the Subcommittee on Energy and Mineral Resources on March 7, 2013, and ordered to be reported by Unanimous Consent on May 15, 2013. The bill would require the Secretary of Interior to conduct an assessment of current global rare earth element resources and the potential future global supply.

H.R. 1063, National Strategic and Critical Minerals Policy Act of 2013

H.R. 1063, the National Strategic and Critical Minerals Policy Act of 2013, was introduced on March 12, 2013, referred to the Subcommittee on Energy and Mineral Resources on March 15, 2013, and ordered to be reported by unanimous consent on May 15, 2013. The bill would require the Secretary of the Interior to conduct an assessment of the current and future demands for the minerals critical to United States manufacturing, agricultural competitiveness, economic and national security.

2013 Annual Industrial Capability Report to Congress

The 2013 Annual Industrial Capability Report to Congress can be accessed online, at http://www.acq.osd.mil/mibp/docs/annual_ind_cap_rpt_to_congress-2013.pdf.

Federal Register, Final Rule on the Production of Specialty Metals

DOD has sought public comment regarding the definition of the word "produced" regarding the production of specialty metals. P.L. 111-383 required DOD to seek public comment regarding the application of the word "produce" as applied to the production of specialty metals. On July 24, 2012, DOD published a proposed rule to amend the DFARS to revise the definition of the word "produce." In the proposed revision and in response to public comments, DOD is proposing to remove certain processes—"quenching and tempering"—from armor plate production, and to "expand the application of the other listed technologies, currently restricted just to titanium and titanium alloys, to any special metal that could be formed by such technologies." The public comment period ended on September 24, 2012, and the final rule was issued on March 28, 2013.[14]

BACKGROUND

Original Congressional Intent

Beginning with the 1973 Defense Appropriations Act, some Members of Congress advocated for a thoughtful and reasonable approach in adding specialty metal to the list of items covered under the Berry Amendment. In the ensuing debate over specialty metal, Senator Jacob Javits discussed the intent of the legislation:

> As an example, I would certainly hope that the Department of Defense in administering this provision would take into consideration the fact that it would be a virtual impossibility for a company participating in a defense contract to try to ascertain for itself, let alone for the myriad of suppliers of small component metals parts, that there was no small amount of metals used which would come within the definition of specialty metals. I would hope that the Department of Defense in the administration of this provision, while seeking to carry out the broad intent of protecting the special metals industry, would have sufficient flexibility and discretion under this provision so that they would not be required to go to ridiculous extremes which would result in an almost impossible administrative burden placed upon Government contractors, and the addition of needless expenses to the Government in carrying out its procurement practices.[15]

The Test of Reasonableness

From the inception of the specialty metal clause, both Congress and DOD emphasized that a test of reasonableness would be applied so that the specialty metal clause would not pose an administrative burden upon DOD contractors or the federal government.

In a March 7, 2007, hearing before the House Armed Services Committee, Air and Land Forces Subcommittee, Lieutenant General Donald J. Hoffman,

Air Force military deputy, asked that Congress give its support to relieving the Air Force from the more arduous aspects of the specialty metal waiver process, as discussed below:

> I would ask for the Committee's continued help in one area, and that is the area of specialty metals. In last year's authorization act, Congress provided some relief in the area of electronic components, where the source of minute amounts of specialty metals cannot be traced throughout the commercial production supply chain. This relief is certainly helpful, but I would ask that there be further consideration for relief in the area of commercial products. Tracing the source of metals and commercial products is very problematic for industry, particularly where DOD is a very small part of their market. The cost of creating a separate supply chain that is able to trace specialty metals down to the lowest tier, such as fasteners, is something industry has been unwilling to accept if it is to remain commercially competitive.
>
> While the Congress has authorized a waiver process, the justification and support of the waivers can be very labor intensive. As an example, the waiver process last year for the AMRAM (Advanced Medium Range Air-to-Air) missile, the government contractor spent over 2,200 man hours to review 4,000 parts, and produced a documentation to justify the waiver. This documentation was eight inches tall in printed form. All this work was to justify a waiver for $14,000 on an item that is valued at $566,000.[16]

Oversight Issues for Congress

The specialty metal clause prohibits DOD from procuring metal that is not produced in the United States.[17] Three issues stand out as policy questions that Congress may choose to consider in its oversight role. First, how does the specialty metal clause affect competition among the different contractor tiers in the U.S. defense industrial base? Second, what are the factors that contribute to the success or failure of the administration of the specialty metal clause? Third, how does one weigh the reliability of having a domestic supplier base in times of urgent and compelling need, coupled with the desire to promote global trade?

Congress may want to consider whether recent revisions in the specialty metal clause have adequately addressed the purpose of the law, as well as concerns raised about how the new rules are implemented.

Competition

Effect on the Defense Industrial Base

Competition for defense work is affected by the availability of sufficient quantity and quality of specialty metal; such metal may be critical and vital to the war-fighting effort if it is used for "high-tech" electronics and communications like personal digital assistants. Creating separate electronic chips for military use only, with no foreign content, would be an expensive undertaking, and some companies have elected not to do so even if it means not being able to sell to DOD.

Congress addressed this issue by exempting electronic components from the specialty metals clause.

> ... unless the Secretary of Defense, upon the recommendation of the Strategic Materials Protection Board pursuant to 10 U.S.C. 187, determines that the domestic availability of a particular electronic component is critical to national security.[18]

Competition Affects Suppliers Differently

Some defense suppliers have framed the specialty metals debate as one between companies that advocate for global trade versus those that favor a dedicated domestic industrial supplier base. On the one hand, some argue that major aerospace companies are eager to seek waivers of domestic source restrictions because doing so would increase their access to foreign markets for specialty metal. Some industry leaders have maintained a view that domestic source restrictions like the Berry Amendment are inconsistent with a policy to encourage global competition. Yet some others believe that the presence of competition, particularly from the foreign markets, makes it more difficult for domestic suppliers to survive. Each supplier in the defense industrial base views competition differently.

For example, some major defense contractors have contended that global competition for commercial and defense work requires establishing and developing foreign trading partners, and that the capacity of domestic suppliers to meet the needs of major defense contractors is insufficient. Some contractors, especially those whose primary market is the U.S. defense industry, know their client base and what they have to buy, and thus are locked into one dedicated supply chain. Yet many other contractors, particularly at the third and fourth tiers of the supply chain, market to both military and commercial sectors; they find that carrying separate supply chains is cost-prohibitive and poses a significant administrative burden. Some companies may not know who the ultimate purchaser of their product will be, so they cannot be certain whether the end use is for a commercial or military application.

The specialty metals clause addresses these concerns: (1) specialty metal from many allied countries is already exempt from the domestic source requirement because of the existence of Memoranda of Understanding (MOU) with the United States, which guarantee that allied suppliers be given equal treatment in DOD procurement;[19] (2) commercial-off-the-shelf (COTS) items are exempt from the specialty metal clause so that suppliers who sell identical commercial and defense articles are exempt from the specialty metal clause; (3) recent changes to the special metals clause allow producers of commercial derivative defense articles and fasteners to treat domestic and foreign specialty metals as fungible materials so that commercial and defense items may be produced on the same production line without the need to trace the source of the metal used in each item.

Congress may want to consider whether recent revisions in the specialty metal clause have adequately addressed this issue, and identify and monitor the impact of this provision within DOD and the DOD supply chain.

Doing Business with DOD Could Mean Maintaining Separate Production Lines

The specialty metal clause as contained in the Berry Amendment required 100% compliance; there was no clause for non-compliant metal. As an example, when DOD purchased avionics, electronics, components, and subassemblies, items with specialty metal were required to be 100% domestic. However, the integration of the global supply chain meant that cheaper, foreign metal could make up virtually all products, and that there were fewer companies that could certify that all of the metal used in the production of their items was wholly domestic in origin. Suppliers who wanted to sell to DOD and to the commercial sector could be forced to maintain two separate production lines; this would raise DOD's costs.

Congress addressed this issue by exempting electronic components from the specialty metal clause, exempting any article containing "de minimis" amounts of specialty metals that were not melted in the United States (de minimis is defined as when non-compliant specialty metal is less than 2% of the total weight of specialty metals in the item). As mentioned earlier, producers of commercial derivative defense items like fasteners can treat domestic and foreign specialty metals as fungible materials so that commercial and defense items may be produced on the same production line without the need to trace the source of the metal used in each item.

Administration

Administration

Can DOD administer and properly execute the new specialty metal clause? The new specialty metal clause may not be entirely enforceable, because it may be nearly impossible to determine to any degree of certainty whether the smallest of the nuts, bolts, screws, and fasteners that make up DOD weapons systems and equipment are of 100% domestic content. Given the recent revisions in the exceptions for electronic components, COTS items, and items with de minimis amounts of non-compliant metal, and the fungible use of specialty metals in commercial derivative articles, items, and fasteners, it may prove difficult for Congress to assess how effectively the specialty metals clause can be administered. Such an assessment would in all likelihood require that DOD collect data on the use of specialty metals under the new rules.

Waivers

The debate over the specialty metal clause was largely fueled by voluntary disclosures, made by companies who sell to DOD, which the companies were in violation of the Berry Amendment specialty metal requirement. For example, the National Semiconductor Corporation disclosed that specialty metal used in its products did not meet the requirement. "To the best of our knowledge, no other semiconductor manufacturer currently is capable of meeting that standard," then wrote Gerry Fields, vice president; Texas Instruments and the Intel Corporation made similar disclosures. Each company stated that, due to the global supply chain for its production line, it would be unable to meet present and future specialty metal requirements.[20] The Semiconductor Industry Association (SIA), which represents about 85% of U.S.-based semiconductor industry, stated that integrated circuits from products made by SIA member companies may contain small quantities of non-compliant specialty metal. Such quantities constitute a small percentage of the item's overall metal content. Further, SIA opined that the application of a domestic preference to specialty metal, as currently applied by DOD and the FY2007 Defense Authorization bill, did not take into account the economic realities that have shaped the development of the specialty metal industry and indeed the entire global technology sector.[21]

During FY2007, DOD approved a "Domestic Non-Availability Determination (DNAD)" to permit the procurement of non-compliant (non-domestic) fasteners.[22] As several suppliers voluntarily disclosed their use of non-compliant specialty metal in defense weapon systems, DOD proposed a temporary modification to the specialty metal clause through a series of interim instructions. On March 10, 2006, the Defense Contract Management Agency issued guidance to its contracting officials on how to handle the acceptance of non-compliant specialty metal, until a long-term solution could be developed. On June 1, 2006, the Under Secretary of Defense for Acquisition, Technology, and Logistics issued a memorandum which authorized a "conditional acceptance and withholding of payment" based on two considerations: (1) a financial consideration (or offset to the federal government) to support the conditional acceptance, and (2) a comprehensive corrective action plan provided by the contractor.[23]

Congress may want to consider whether to carefully monitor the waiver process to see if the need for waivers has been rendered unnecessary due to recent changes in the law.

Effect on Joint Ventures and Partnerships

There are contrasting views on the effect of the specialty metal clause on joint ventures and partnerships. One view was expressed by some companies that signaled their inability to meet the specialty metal requirement. They were part of the Berry Amendment Reform Coalition, an organization of industry associations that represents thousands of companies that provide products, services, and personnel to the federal government. The coalition asserted that the specialty metal clause could have a harmful effect on the ability of defense contractors to partner with other companies. Prime contractors who relied on small and mid-size companies to deliver components, such as fasteners and components from electronic circuit boards, found compliance with the Berry Amendment difficult. According to the coalition, due to Berry Amendment requirements, the cost of a fastener for a military plane can be as much as five times more than the cost of a fastener for a commercial airplane. Additionally, the cost of using domestic titanium (for a U.S. company) could be as much as 40% higher than the cost of using non-domestic titanium.[24]

Another view was expressed in a 2009 report prepared by the RAND Project Air Force (PAF),[25] which studied factors that affected price fluctuations in titanium. The report concluded that there are a number of market forces and price drivers that affect the domestic titanium industry.[26]

The Administrative Burden

The cost of compliance with administrative requirements of the specialty metal clause could be unsustainable. Many companies report that they are unable to develop a compliance measure that would support a 100% across the board systematic reporting system of every type of metal that is used in the melting process. Such a system of compliance would be difficult, if not impossible, to maintain. Further, since most contractors have smaller percentages of their business line devoted to DOD contracts, it may not be cost-effective for contractors to develop such a system. Many have signaled that if forced to do so, they would terminate their business relationship with DOD and increase their capacity for commercial work.

Congress may want to determine whether the recent amendments to the specialty metals law have substantially mitigated any possible administrative burdens.

Reliability

In Urgent Situations and Times of War

The issue of reliability has been the cornerstone of why domestic source restrictions, like the specialty metal clause, are viewed by some proponents as essential to the viability of the domestic defense industrial base. Central to the issue of reliability is the basic premise upon which the Berry Amendment was first adopted. The Berry Amendment, which dates from the eve of World War II, was established for a narrowly defined purpose: to ensure that U.S. troops wore military uniforms wholly produced within the United States and to ensure that U.S. troops were fed with food products solely produced in the United States.[27] There were at least two congressional concerns: (1) that the United States maintain a vibrant domestic industrial base by requiring that military troops wear uniforms made in the United States, and consume food produced in the United States; and (2) that the nation be prepared in the event of adversity or war.

Many view domestic source restrictions like the specialty metal clause as a way to insure that, in urgent situations and times of war, the United States will have access to critical items needed to ensure national security. Those who advocate for maintaining a robust capability among the domestic sources for titanium, as an example, argue that these companies will ensure that, should a global shortage of titanium develop or if the United States loses a key trading partner, the United States will not become unduly dependent on another country for a critical item. Furthermore, having domestic suppliers who have the protection of the specialty metal clause may ensure that domestic production lines remaining open and viable.

Maintaining a Productive and Profitable Domestic Base

An argument that is often raised is that, as an example, the three domestic titanium producers would not be viable if the specialty metal clause did not exist. A look at the three domestic titanium producers reveals that they have different income streams, and are not wholly dependent on the specialty metals clause.

Domestic Restrictions Protect the U.S. Industrial Base

Some policymakers believe that products consumed by Americans should be made at home, and that domestic source provisions like the specialty metal clause represent jobs for the smaller, domestic companies in America. However, when compared to the jobs generated by the major defense contractors in the global supply chain, the number of local jobs is proportionately smaller.

OPTIONS FOR CONGRESS

It is important to note that the specialty metal clause in the Berry Amendment had been in place for over three decades. Any change in the law could likely have both upstream and downstream effects in the defense supply chain. How will the change affect prime contractors and subcontractors on the second, third, and fourth tiers, as well as domestic suppliers?

At least seven possible options for policymakers to consider are listed below: (1) eliminate the specialty metal clause; (2) require an assessment of compliance with new exceptions to the specialty metals clause; (3) require a review of waivers issued under the revised specialty metals clause, including requiring DOD to publicly disclose when waivers are granted; (4) require congressional approval before non-compliant specialty metal can be used in certain defense contracts; (5) require a congressional report for each platform/component where non-compliant specialty metals are used in defense contracts; (6) encourage the use of domestic specialty metal; and (7) appoint a special metals protection board.

Eliminate the Specialty Metal Clause

Congress could eliminate the specialty metal clause as well as the Berry Amendment. Some question whether the Berry Amendment is still a good policy, given the global supply chain; others question whether each item needs the protection of a domestic source policy. Possibly the elimination of the specialty metal clause or the Berry Amendment would be met with fierce opposition, particularly from domestic suppliers without a strong foreign market. To some extent, domestic source restrictions like the Berry Amendment may help to insure that there is a dedicated domestic source for DOD products.

Require an Assessment of Compliance with New Exceptions to the Specialty Metal Clause

Congress could require a comprehensive review of compliance and implementation of the revisions to the new specialty metal clause. Congress could require that GAO complete a study of the frequency and impact of the waiver process, use of exceptions (such as the exceptions for national security, de minimis, and COTS) and evaluate DOD's role in exercising oversight of the implementation of the new provisions.

Require a Review of Waivers Issued Under the Revised Specialty Metal Clause, and for Public Disclosure When Waivers Are Granted

Congress could require a review and publication of the numbers and types of waivers granted (Determinations of Domestic Non-Availability [DNAD])28 to purchase items that are noncompliant, and tighten the waiver process so that waivers are not granted for inappropriate or arbitrary reasons. Congress could direct DOD to work more closely with the Defense Contract Management Agency to determine if there is greater compliance with the specialty metal clause. On the one hand, requiring more transparency and openness in the waiver process may pose more of an administrative burden on DOD. On the other hand, more transparency may engender more public confidence in the process.

Require Congressional Approval before Non-Compliant Specialty Metal Can Be Used In Defense Contracts

Congress could limit the use of non-compliant specialty metal. One approach is the application of a market-based standard—so that DOD can tie the amount of non-compliant specialty metal permitted to the percentage of business that the contractor has with DOD—so if a contractor acquires 16% of the DOD market, it will be permitted to use compliant specialty metal for at least 16% of its total market needs.

As an example set forth in 10 U.S.C. 2306(b), Congress enacted six legal criteria that must be met for the Multi-Year Procurement Program (MYP) to be operational, and has reached conclusions as to when a weapons program does not meet all of the requirements for MVP.[29] The process has been perceived, for the most part, as fair and balanced.

Such a set of conditions could determine under what circumstances non-compliant specialty metal could be used in defense contracts, and might include the following criteria:

- That the use of non-compliant specialty metal will result in a substantial savings of the total anticipated costs throughout the life of the contract;
- That the percentage of specialty metal used for the weapons program is expected to remain substantially unchanged during the contract period, in terms of rate of production and procurement, and total quantities;
- That the contract for the use of non-compliant specialty metal will be subject to re-competition on a five-year basis, to give the domestic specialty metal industry an opportunity to develop the capacity and capability to meet future program requirements;
- That the estimates of both the cost of the contract and the anticipated cost avoidance are realistic and supportable through independent audits and investigations;

- That there is a reasonable expectation that throughout the life of the contemplated contract period, the head of the military service will request funding for the contract, at the level required, to avoid contract cancellation; and
- That the use of such non-compliant specialty metal, in this particular weapons system, is critical to the national security of the United States.

Require a Congressional Report for Each Platform/Component Where Foreign Specialty Metals are used in Defense Contracts

One approach that Congress may consider is to require DOD to produce a separate report for each platform or component of a weapons program where foreign specialty metals are used. For example, in the Future Combat System, where there are about 20-24 separate platforms, each platform would be supported by a separate report which calculates the sources, types, and percentages of specialty metal content, both foreign and domestic.

Examining the specialty metal content by platform may uncover data that are often buried in the aggregate numbers of larger reports on the entire weapons program. The level and specificity of detail could pose an administrative burden on DOD and defense contractors.

Encourage the Use of Domestic Specialty Metal

Congress could develop steps to further support a stronger domestic specialty metal industry. One alternative is to encourage the development of technological capabilities and advances by providing tax incentives for investment in scientific and manufacturing technology. Congress could create a socioeconomic subsidy program to support the domestic specialty metal suppliers; one approach would be to create a partnership between DOD and domestic suppliers. Such an approach was described as a way to develop a greater capacity to meet the delivery requirements for aviation parts in the military, as noted in the FY2007 National Defense Appropriations Act, P.L. 109-289, where the increased demand for domestic steel suppliers was highlighted:

> The Department of Defense's demand for iron-based alloy aviation specialty steels has dramatically increased as a result of continuing

deployments to the overseas theaters of operation. Today, there is only one domestic supplier for a unique process which utilizes vacuum inducted melt/vacuum arc re-melt, the process which gives aviation grade steels their required properties. These specialty steels are critical to building high technology U.S. military weapon systems. Further, there has been a related and dramatic increase in the raw material needed to make these specialty steels. Lead times for these raw materials have grown from 3 months to 1 year. According to the Army, the overall effect on lead times for spare part deliverables has swelled in some cases to greater than 24 months. As such, the conferees encourage the Department of Defense to partner with domestic industry to develop a greater capacity to meet the delivery requirements for aviation parts to the military within an acceptable time frame. The conferees suggest that the Department explore a 50/50 cost share project between the Federal government, private industry, and state governments as the best means to create this capacity as rapidly as possible.[30]

Appoint a New Specialty Metals Protection Board

The House of Representatives has expressed concern over the conclusions expressed by the Specialty Metals Protection Board, particularly over the Board's definition of what constitutes a critical and strategic material. In the House Armed Services Committee report accompanying both the FY2009 and FY2010 National Defense Authorization Acts, it was noted that the conclusions reached by the Board were unacceptable. Excerpts from the House report accompanying H.R. 2647, the FY2010 National Defense Authorization Act, appear below.[31]

> This definition limits the purview of the Board to only those materials for which the determinations the Board is tasked to make are presupposed in the definition of the materials themselves. Furthermore, such a definition fails to include a range of materials that Congress has designated as critical to national security and, as such, has provided significant protection or domestic preference in DOD policy and in statute. For example, Congress has determined that reliance on foreign sources of supply for materials such as titanium, specialty steel, and high performance magnets, poses a heightened risk. The Board's narrowing of the definition of materials critical to national security renders the Board unable to provide perspective on the adequacy, suitability, or

> effectiveness of those policies. Moreover, it limits the ability of the Board to consider any course of action, however minor, in relation to a material until the point at which potential damage to national security is imminent and severe. It also creates the perverse situation that a material could be critical to every element of the industrial base upon which the Department depends, but not considered critical to the Department itself if the material is also used significantly in commercial items. As an indication of the inadequacy of this definition for the Board's functioning, the Board currently identifies only one material as meeting the definition for consideration as a strategic material critical to national security. The committee does not find this conclusion to be plausible and expects that the Board will swiftly revisit this definition to ensure that it is able to identify gaps in our domestic defense supply chain and provide the President, the Secretary of Defense, and Congress with information, analysis, and advice on strategic materials which are critical to the operations of the Department of Defense.

Congress could follow the example of the Packard Commission by creating an independent body to study the specialty metal clause and its impact on the defense industry.[32] By requiring an independent review chartered by Congress, industry experts could be consulted to determine if DOD's determination that only one material is "critical" to national security and that specialty metals do not require domestic preference are reasonable.

LEGISLATION

Legislation Passed in the 112th Congress

H.R. 4310, the National Defense Authorization Act for FY2013, contains a provision which broaden the definition of specialty metals produced within the United States. Section 817 would amend 10 U.S.C. 2533b to include "melted, or processed in a manner that results in physical or chemical property changes that are the equivalent of melting." The term "produced" does not include finishing processes such as "rolling, heat treatment, quenching, tempering, grinding, or shaving."[33]

H.R. 4310 also contains a provision, Section 901, which reconfigures the membership and scope of the Strategic Materials Protection Board (SMPB) by requiring that the Deputy Assistant Secretary of Defense for Manufacturing and Industrial Base Policy serve as chair. The congressional intent of such a restructuring can be found in a statement of findings that accompanies the bill. Excerpts from the bill follow.

TITLE IX—DEPARTMENT OF DEFENSE ORGANIZATION AND MANAGEMENT Subtitle A—Department of Defense Management

SEC. 901. ADDITIONAL DUTIES OF DEPUTY ASSISTANT SECRETARY OF DEFENSE FOR MANUFACTURING AND INDUSTRIAL BASE POLICY AND AMENDMENTS TO STRATEGIC MATERIALS PROTECTION BOARD.

(a) Findings- Congress finds the following:

(1) The Defense Logistics Agency has made little progress in addressing the findings and recommendations from the April 2009 report of the Department of Defense report titled `Reconfiguration of the National Defense Stockpile Report to Congress'.
(2) The office of the Deputy Assistant Secretary of Defense for Manufacturing and Industrial Base Policy has historically analyzed the United States defense industrial base from the point of view of prime contractors and original equipment manufacturers and has provided insufficient attention to producers of materials critical to national security, including raw materials producers.
(3) Responsibility for the secure supply of materials critical to national security, which supports the defense industrial base, is decentralized throughout the Department of Defense.
(4) The office of the Deputy Assistant Secretary of Defense for Manufacturing and Industrial Base Policy should expand its focus to consider both a top-down view of the supply chain, beginning with prime contractors, and a bottom-up view that begins with raw materials suppliers.
(5) To enable this focus and support a more coherent, comprehensive strategy as it pertains to materials critical to national security, the office of the Deputy Assistant Secretary of Defense for Manufacturing and Industrial Base Policy should develop policy, conduct oversight, and monitor resource allocation for agencies of the Department of Defense, including the Defense Logistics Agency, for all activities that pertain to ensuring a secure supply of materials critical to national security.
(6) The Strategic Materials Protection Board should be reconfigured so as to be chaired by the Deputy Assistant Secretary of Defense for Manufacturing and Industrial Base Policy and should fully execute its duties and responsibilities.[34]

H.R. 3449, Defense Supply Chain and Industrial Base Security Act was introduced on November 16, 2011, and referred to the House Armed Services Committee. The bill would require the Secretary of Defense to

(1) develop a defense supply chain and industrial base strategy, and subsequent plan, designed to secure the supply chain and industrial base sectors determined to be critical to U.S. national security;
(2) review the strategy and plan on a biennial basis; and
(3) report to Congress on the strategy and plan, as well as on the results of each review.[35]

FY2008 National Defense Authorization Act and Revisions to Existing Specialty Metal Rules

P.L. 110-181, the FY2008 National Defense Authorization Act, contained three new provisions which affect the specialty metal clause:

- Section 803 required the Strategic Materials Protection Board to perform an assessment of the viability of domestic producers of strategic materials, the purpose of which is to assess which domestic producers are investing, or plan to invest on a sustained basis, in the development of a continued domestic production capability of strategic materials to meet national defense requirements. Such an assessment would be evaluated and weighted in any decision to grant future waivers to the specialty metal clause.[36]

Sections 804 and 884 included these provisions:

- Applies to contracts or subcontracts for the acquisition of specialty metals, including mill products, such as bar, billet, slab, wire, plate and sheet, that have not been incorporated into end items, subsystems, assemblies, or components;
- Applies to contracts or subcontracts for the acquisition of forgings or castings of specialty metals, unless such forgings or castings are incorporated into commercially available off-the-shelf end items, subsystems, or assemblies;
- Applies to contracts or subcontracts for commercially available, high performance magnets unless such high performance magnets are

incorporated into commercially available, off-the-shelf-end items or subsystems;

- Applies to contracts or subcontracts for commercially available off-the-shelf fasteners (some exceptions noted);
- Applies to contracts for electronic components unless the Secretary of Defense, upon the recommendation of the Strategic Materials Protection Board pursuant to Section 187 of this title, determines that the domestic availability of a particular electronic component is critical to national security;
- Provides exceptions for the Secretary of Defense or the Secretary of a military department to accept delivery of an item containing specialty metals that were not melted in the United States, if the total amount of noncompliant specialty metals in the item does not exceed 2% of the total weight of specialty metals in the item (excludes high performance magnets);
- Provides exceptions for the Secretary of Defense or the Secretary of a military department to acquire commercial derivative military articles, under certain conditions; and
- Provides waivers for national security if the Secretary of Defense determines in writing that acceptance of such end items is necessary to the national security interests of the United States, under certain conditions.

FY2007 National Defense Authorization Act and the Enactment of a New Specialty Metal Clause

Congress enacted provisions in the FY2007 National Defense Authorization Act that changed the statutory authority for the special metal clause. P.L. 109-364 created a new specialty metal clause separate from the Berry Amendment, moving it into a separate section of Title 10.[37] Also, Section 842b established a one-time waiver of the Berry Amendment for non-compliant specialty metal incorporated into items produced, manufactured, or assembled in the United States before the date of the act's enactment. DOD can grant waivers provided the noncompliance was not knowing or willful.[38]

Section 842 of the FY2007 National Defense Authorization Act

- Creates a new section, 10 U.S.C. 2533b, in the U.S. Code;

- Reaffirms that any specialty metal (e.g., raw stock) acquired directly by the government or by a prime contractor for delivery to the government, must be melted or produced in the United States. This restriction applies to all tiers of subcontractors when acquiring aircraft, missile and space systems, ships, tank and automotive items, weapons systems, or ammunition;
- Restricts not only the procurement of the specialty metal in these items, but restricts procurement of the end items, and components thereof, that contain specialty metals;
- Prohibits the practice of withholding contract payment while conditionally accepting noncompliant items in these categories;
- Provides a national security exemption, where the Secretary of Defense or the Secretary of the military service can waive the specialty metal compliance requirement when compliant specialty metal, of a satisfactory quality, quantity, and in the required form cannot be procured as needed and when needed;
- Exempts procurement outside the United States in support of combat or contingency operation;
- Exempts sole source procurements based on unusual and compelling urgency of need;
- Provides exceptions for procurements related to international agreements (e.g., qualifying countries), non-appropriated fund contracts, and small purchases (i.e., less than the simplified acquisition threshold); and
- Provides for a "de minimis" exception which means that up to 2% of noncompliant specialty metal can be included in electronic components.[39]

Strategic Materials Protection Board

Section 843 of FY2007 National Defense Authorization Act required the creation of a Strategic Materials Protection Board to determine which items should be designated as critical to national security, and to recommend changes for future domestic sourcing.

2007 Meeting of the Strategic Materials Protection Board

The board held its first meeting on July 17, 2007, and reached the following conclusions as described here:

- the term "materials critical to national security" would be taken to mean "strategic materials critical to national security" or simply "strategic materials," and would include those specialty metals listed in 10 U.S.C. 2533b, and any other materials that the board chose to so designate;
- the board's efforts would initially focus on determining the need to take action to ensure a long-term domestic supply of specialty metals as designated in 10 U.S.C. 2533b;
- the board adopted certain Terms of Reference to shape its deliberations; and
- the board directed its executive secretary to conduct an initial analysis of national security issues associated with strategic materials, and to report the results of that analysis at the next SMPB meeting.[40]

2008 Meeting of the Strategic Material Protection Board

The board held its second meeting on December 12, 2008, and reached the following conclusions as described here:

- Approved a definition of "strategic material" and "critical material" proposed by the executive secretary were discussed and approved by the board;
- Reviewed and validated the work of the Strategic and Critical Materials Working Group in response to congressionally directed requirements of H.Rept. 109-89 and S.Rept. 110-55;
- Validated an "Initial Analysis of National Security Issues Associated with Strategic Materials" and authorized its publication in the Federal Register; and
- Revised the Terms of Reference to reflect their new definitions for strategic and critical materials, providing the board with more flexibility to examining future issues, and broadening their scope to address additional matters associated with strategic materials.[41]

Based on the recommendations of the December 2008 meeting of the Strategic Materials Protection Board, DOD has determined that specialty metals no longer require domestic source protection, as described below in excerpts from the report to Congress.

> The key finding of this analysis is that specialty metals, as defined in 10 U.S.C. 2533b, are not "materials critical to national security" for which only a U.S. source should be used; and there is no national security reason for the Department to take action to ensure a long term domestic supply of these specialty metals. The "criticality" of a material is a function of its importance in DOD applications, the extent to which DOD actions are required to shape and sustain the market, and the impact and likelihood of supply disruption. The analysis showed that specialty metals are "strategic materials" which may require special monitoring and attention/action; but not, in general, a domestic source restriction. Should reliable supplies/capacities be insufficient to meet potential requirements for a projected conflict, other risk mitigation options, including stockpiling, could represent an effective alternative.
>
> High purity beryllium, however, is a critical material. Even in peacetime, defense applications dominate the market; it is essential for important defense systems and unique in the function it performs. In addition, domestic production capabilities have atrophied, and there are no reliable foreign suppliers. Accordingly, the Department should continue to take those special actions necessary to maintain a long term domestic supply of high purity beryllium. In fact, the Department has established a Title III of the Defense Production Act project with U.S. supplier Brush-Wellman to build and operate a new high purity beryllium production facility.
>
> The Strategic Materials Protection Board (SMPB) should review and validate any internal or external recommendations that identify strategic materials that are essential for a wide variety of important defense applications and for which there is a relatively high potential for supply disruption. For example, a relatively high potential for supply disruption would be represented by a situation in which reliable supplies (U.S. or nonU.S.) are projected to be insufficient to support the defense needs of the United States during peacetime and/or during a conflict. In such circumstances, DOD market intervention such as increasing or establishing reliable production capability and/or stockpiling may be an effective risk mitigation strategy.[42]

Annual Industrial Capabilities Report to Congress

Title 10 of the United States Code, Section 2504, requires the Secretary of Defense to report to the House and Senate Armed Services Committees on the viability of the defense industrial base including the following information:

(1) A description of the departmental guidance prepared pursuant to section 2506 of this Title.
(2) A description of the methods and analyses being undertaken by the Department of Defense alone or in cooperation with other Federal agencies, to identify and address concerns regarding technological and industrial capabilities of the national technology and industrial base.
(3) A description of the assessments prepared pursuant to section 2505 of this Title and other analyses used in developing the budget submission of the Department of Defense for the next fiscal year.
(4) Identification of each program designed to sustain specific essential technological and industrial capabilities and processes of the national technology and industrial base."

2012 Report to Congress

The 2012 Annual Industrial Capabilities Report to Congress can be accessed online.[43]

End Notes

[1] The specialty metal clause of the Berry Amendment was enacted in the 1973 DOD Appropriations Act, P.L. 92-570. For more information on the Berry Amendment, see CRS Report RL31236, The Berry Amendment: Requiring Defense Procurement to Come from Domestic Sources, by Valerie Bailey Grasso.

[2] P.L. 92-570, the 1973 DOD Appropriations Act, was amended to add the following text: "Wood (whether in the form of fiber or yarn or contained in fabrics, materials, or manufactured articles), or specialty metals not grown, reprocessed, reused, or produced in the United States or its possessions."

[3] The specialty metal provision of the Berry Amendment was enacted in the 1973 DOD Appropriations Act, P.L. 92-570. For more information on the Berry Amendment, see CRS Report RL31236, The Berry Amendment: Requiring Defense Procurement to Come from Domestic Sources, by Valerie Bailey Grasso.

[4] For the latest DOD information on specialty metals, see Defense Logistics Agency, Information on Specialty Metals, http://www.dla.mil/J-7/metals.asp#changes.

[5] DFAR 225.252.7009 Restriction on the Acquisition of Certain Articles Containing Specialty Metals, at http://www.acq.osd.mil/dpap/dars/dfars/pdf/r20120420/252225.pdf.

[6] The qualifying countries are Austria, Belgium, Canada, Denmark, Egypt, Germany, France, Greece, Israel, Italy, Luxembourg, Netherlands, Norway, Portugal, Spain, Sweden, Switzerland, Turkey, United Kingdom and Northern Ireland. See DFAR 225.252.7014.

[7] This exception allows producers of commercial derivative defense articles, like fasteners, to treat domestic and foreign specialty metals as fungible, so that commercial and defense articles may be produced on the same production line without the need to trace the source of the metal used in each article.

[8] Title 10, Subtitle A, Part IV, Chapter 148, Subchapter V, 2533b. Requirement to buy strategic materials critical to national security from American sources; exceptions; http://www.acq.osd.mil/dpap/cpic/ic/ restrictions_on_specialty_metals_10_usc_2533b.html.

[9] Public Law (P.L.) 113-66 was signed into law on December 26, 2013.

[10] H.Rept. 113-102, to accompany H.R. 1960, the proposed National Defense Authorization Act for Fiscal Year 2014.

[11] H.Rept. 113-102, to accompany H.R. 1960, the proposed National Defense Authorization Act for Fiscal Year 2014.

[12] S. 1600, the Critical Minerals Policy Act of 2013.

[13] H.R. 4402, the National Strategic and Critical Minerals Production Act of 2012 was introduced on July 16, 2012, and referred to the Committee on Energy and Natural Resources.

[14] DFARS Case 2012-D041. Specialty Metals and the Definition of "Produce." Federal Register, Vol. 78, No. 60, March 28, 2013, at http://www.acq.osd.mil/dpap/dars/dfars/changenotice/2013/20130328/fr_2012-D041.pdf.

[15] Excerpt from Senator Jacob Javits' remarks on the passage of H.R. 16593, Making Appropriations for the Defense Establishment for Fiscal Year 1973. 118 Congressional Record S17967 (October 13, 1972).

[16] Statement of Lieutenant General Donald J. Hoffman, Military Deputy, Office of the Assistant Secretary of the Air Force for Acquisition, before the Subcommittee of Air and Land Forces, House Armed Services Committee, March 7, 2007.

[17] 10 U.S.C. §2533a, Requirement to Buy Certain Articles from American Sources; Exceptions.

[18] 252.225.70-Authorization Acts, Appropriations Acts, and Other Statutory Restrictions on Foreign Acquisition (Revised March 17, 2011), at http://www.acq.osd.mil/dpap/dars/dfars/html/current/225_70.htm#225.7003-3.

[19] See list of qualifying countries at http://www.acq.osd.mil/dpap/dars/dfars/html/ current/252225.htm#252.225-7012. These countries are allies of the United States. Japan is an ally and significant specialty metal producing country but is not listed here, largely attributed to the fact that Japan has a constitutional prohibition on weapons production.

[20] Memorandum on Berry Amendment/Buy American Act - DFARS Clause 252.225-7014, Gerry Fields, Vice-President, Worldwide Quality Network and New Product Execution. National Semiconductor Corporation, March 7, 2006; also, Request for Confirmation of Compliance with the Berry Amendment, by Brent Thornton, Quality Assurance Manager, Hirel, Defense, and Aerospace Products, March 23, 2006; Memorandum on Domestic Preference for Specialty Metals, Texas Instruments, May 12, 2006.

[21] SIA's Position on the Berry Amendment, May 9, 2006. Since 1977, SIA has identified itself as a leading voice for the semiconductor industry. SIA member companies comprise more than 85% of the U.S. semiconductor industry. Collectively, the chip industry employs a domestic workforce of 225,000 people. According to SIA, over 70% of U.S. manufacturing facilities are on U.S. soil, but greater than 75% of the industry revenue is affected by specialty metal provisions. These provisions affect military contracts and the availability of commercial products for the military. SIA points out that procurement regulation affect semiconductors

in two ways: first, the military relies on a commercial offthe-shelf (COTS) acquisition model for many components. Second, semiconductors are used in downstream products supplied under military contracts and subcontracts. Because of these trends, they note their customers have a more direct exposure to government procurement than do semiconductor companies themselves; see http://www.siaonline.org.

[22] http://www.dcma.mil/dnad/. DOD has issued DNADs for a number of items. In the Fastener DNAD, it was determined that satisfactory quality and sufficient quantity of specialty metal in the form of fasteners could not be procured as and when needed. Subsection (b) of 10 U.S.C. 2533b states that if such a determination is made, subsection (a) does not apply. Thus, the restriction in subsection (a) of 10 U.S.C. 2533b does not apply to fasteners. Contracting officers may procure end items, and components thereof, containing fasteners, notwithstanding the country where the specialty metals contained in such items were melted or produced. DOD will revisit the basis for this DNAD if it learns that the circumstances which formed the basis of the determination have changed. Thus, the DNAD will be revised if and when compliant specialty metal of satisfactory quality and sufficient quantity, in the required form, can be procured as and when needed.

[23] Defense Contract Management Agency, Interim Instruction, Non-compliance with the Preference for Domestic Specialty Metals Provision, DFARS 252.225-7014, revised March 10, 2006, 4 p.; and OUSD(A&TL). Memorandum on Berry Amendment Compliance for Specialty Metals, by Kenneth J. Krieg, June 1, 2006, 2 p.

[24] Senate Berry Amendment Streamlining Proposal: Myth versus Reality. A position paper of the Berry Amendment Reform Coalition. July 18, 2006, 4 p. The Berry Amendment Reform Coalition is an organization of about a dozen industry associations that reportedly support alternative approaches that promote a reasonable and balanced solution. http://www.nedassoc.org/.

[25] PAF is a division of the Rand Corporation and the U.S. Air Force's federally funded center for research and analysis. See http://www.rand.org/paf.

[26] Seong, Somi, et al. Titanium: Industrial Base, Price Trends, and Technology Initiatives. RAND Project AIR FORCE, March 29, 2009.

[27] On April 5, 1941, the Berry Amendment was enacted as part of the Fiscal Year (FY) 1941 Fifth Supplemental National Defense Appropriations Act, P.L. 77-29, 10 U.S.C. § 2241 note. The Berry Amendment was made permanent when P.L. 102-396, §9005, was amended by P.L. 103-139, §8005. Since then, Congress has regularly added or subtracted Berry Amendment provisions. On December 13, 2001, passage of the FY2002 National Defense Authorization Act codified the Berry Amendment, repealing §§9005 and 8109 of the above-mentioned bills. The Berry Amendment is now codified at 10 U.S.C. 2533a.

[28] http://www.acq.osd.mil/dpap/cpic/ic/domestic_non-availability_determinations_dnads.html.

[29] According to the Defense Acquisition University, a multi-year procurement (MYP) is "a method of competitively purchasing up to 5 years' requirements in one contract, funded annually as appropriations permit." Congress set up specific rules that must be met before a program gains MYP status.

[30] H.Rept. 109-676.

[31] U.S. Congress, House Armed Services Committee, H.Rept. 111-166, Report on H.R. 2647: National Defense Authorization Act for Fiscal Year 2010, P.L. 111-84, signed into law on October 28, 2009.

[32] In July 1985, President Reagan asked David Packard, chairman of the Hewlett-Packard Corporation and a former Deputy Secretary of Defense, to chair an independent Blue Ribbon Commission which came to be known as the Packard Commission. The Packard

Commission was directed to conduct a broad study of defense management including the budget process, procurement, organization and operation, and legislative oversight, and to make recommendations for streamlining and improving defense management. Executive Order 12526, July 15, 1986.

[33] H.R. 4310 was introduced on March 29, 2012, passed on a recorded vote on May 12, 2013, and referred to the Senate Armed Sources Committee on July 19, 2912.

[34] H.R. 4310, Title 9, Department of Defense Organization and Management, Subtitle A, Department of Defense Management.

[35] H.R. 3449, Bill Summary, Legislative Information Service.

[36] §803. Reinvestment in Domestic Sources of Strategic Materials, P.L. 110-181.

[37] 10 U.S.C. 2533b.

[38] FARS 225.7003-4.

[39] The final rule was published in the Federal Register (FR Doc No: E9-17967, Federal Register, Volume 74, Issue 144, Wednesday, July 29, 2009.) The final rule under DFARS Case 2008–D003 defined "produce" to mean "the application of forces or processes to a specialty metal to create the desired physical properties through quenching or tempering of steel plate, gas atomization or sputtering of titanium, or final consolidation of non-melt derived titanium powder or titanium alloy powder." http://www.gpo.gov/fdsys/pkg/FR-2009-07-29/html/E9-17967.htm.

[40] Office of the Under Secretary of Defense (Acquisition, Technology & Logistics.) First meeting of the Strategic Materials Protection Board, September 2007, p. 2.

[41] Office of the Under Secretary of Defense (Acquisition, Technology & Logistics), Deputy Under Secretary of Defense (Industrial Policy), and the Executive Secretary to the Strategic Materials Protection Board. Report of the Meeting of the Department of Defense Strategic Protection Materials Board. December 12, 2008, pages 1-4.

[42] Office of the Under Secretary of Defense (Acquisition, Technology & Logistics), Deputy Under Secretary of Defense (Industrial Policy), and the Executive Secretary to the Strategic Materials Protection Board. Report of the Meeting of the Department of Defense Strategic Materials Board. December 12, 2008, p. 5-6. A summary of the Board's analysis and conclusions were published in the Federal Register, Vol. 74. No. 34, February 23, 2009, pages 8061-8064.

[43] Department of Defense. Office of Under Secretary of Defense for Acquisition, Technology & Logistics Industrial Policy. Annual Industrial Capabilities Report to Congress, May 2010, at http://www.acq.osd.mil/mibp/docs/annual_ind_cap_rpt_to_congress-2012.pdf.

Index

A

B

C

H

I

J

L

M

N

O